Lecture Notes in Computer Science 8999

Commenced Publication in 1973
Founding and Former Series Editors:
Gerhard Goos, Juris Hartmanis, and Jan van Leeuwen

T0211694

Wanming Chu   Shinji Kikuchi
Subhash Bhalla (Eds.)

Databases in Networked Information Systems

10th International Workshop, DNIS 2015
Aizu-Wakamatsu, Japan, March 23-25, 2015
Proceedings

 Springer

Volume Editors

Wanming Chu
Shinji Kikuchi
Subhash Bhalla

University of Aizu, Database Systems Laboratory
Aizu-Wakamatsu, Fukushima 965-8580, Japan
E-mail: w-chu@u-aizu.ac.jp
        shinji.kikuchi@kne.biglobe.ne.jp
        bhalla@u-aizu.ac.jp

ISSN 0302-9743   e-ISSN 1611-3349
ISBN 978-3-319-16312-3   e-ISBN 978-3-319-16313-0
DOI 10.1007/978-3-319-16313-0
Springer Cham Heidelberg New York Dordrecht London

Library of Congress Control Number: 2015932993

LNCS Sublibrary: SL 3 – Information Systems and Application, incl. Internet/Web
and HCI

Typesetting: Camera-ready by author, data conversion by Scientific Publishing Services, Chennai, India

Printed on acid-free paper

Springer is part of Springer Science+Business Media (www.springer.com)

Preface

Business data analytics in scientific domains depends on computing infrastructure. A scientific exploration of data is beneficial for large-scale public utility services, either directly or indirectly. Many research efforts are being made in diverse areas, such as big data analytics and cloud computing, sensor networks and high-level user interfaces for information accesses by common users. Government agencies in many countries plan to launch facilities in education, health-care, and information support as a part of e-government initiatives. In this context, information interchange management has become an active research field. A number of new opportunities have evolved in design and modeling based on the new computing needs of the users. Database systems play a central role in supporting networked information systems for access and storage management aspects.

The 10th International Workshop on Databases in Networked Information Systems (DNIS) 2015 was held during March 23–25, 2015 at University of Aizu in Japan. The workshop program included research contributions, and invited contributions. A view of research activity in information interchange management and related research issues was provided by the sessions on related topics. The keynote address was contributed by Prof. Divyakant Agrawal. The session on "Information and Knowledge Management" had an invited contribution from Dr. Paolo Bottoni. The following section on "Business Data Analytics and Visualization," had an invited contribution from Dr. Arnab Nandi. The session on "Networked Information Resources" includesd an invited contribution by Dr. Shelly Sachdeav. The section on "Business Data Analytics in Astronomy and Sciences," had an invited contribution by Dr. Naoki Yoshida. I would like to thank the members of the Program Committee for their support and all authors who considered DNIS 2015 in making research contributions.

The sponsoring organizations and the Steering Committee deserve praise for the support they provided. A number of individuals contributed to the success of the workshop. I thank Dr. Umeshwar Dayal, Prof. J. Biskup, Prof. D. Agrawal, Dr. Cyrus Shahabi, Prof. T. Nishida, and Prof. Shrinivas Kulkarni for providing continuous support and encouragement.

The workshop received invaluable support from the University of Aizu. In this context, I thank Prof. Ryuichi Oka, President of University of Aizu. Many thanks also to the faculty members at the university for their cooperation and support.

March 2015

Wanming Chu
Shinji Kikuchi
Subhash Bhalla

Organization

The DNIS 2015 international workshop was organized by the Graduate Department of Information Technology and Project Management, University of Aizu, Aizu-Wakamatsu, Fukushima, Japan.

Steering Committee

Divyakant Agrawal	University of California, USA
Hosagrahar V. Jagadish	University of Michigan, USA
Masaru Kitsuregawa	University of Tokyo, Japan
Toyoaki Nishida	Kyoto University, Japan
Krithi Ramamritham	Indian Institute of Technology, Bombay, India
Cyrus Shahabi	University of Southern California, USA

Executive Chair

N. Bianchi-Berthouze	University College London, UK

Program Chair

S. Bhalla	University of Aizu, Japan

Publicity Committee Chair

Shinji Kikuchi	University of Aizu, Japan

Publications Committee Co-chair

Wanming Chu	University of Aizu

Program Committee

D. Agrawal	University of California, USA
V. Bhatnagar	University of Delhi, India
P. Bottoni	La Sapienza University of Rome, Italy
L. Capretz	University of Western Untario, Canada
Richard Chbeir	Bourgogne University, France
G. Cong	Nanyang Technological University, Singapore
Pratul Dublish	Microsoft Research, USA
Fernando Ferri	IRPPS - CNR, Rome , Italy
W.I. Grosky	University of Michigan-Dearborn, USA
J. Herder	University of Applied Sciences, Fachhochschule Düsseldorf, Germany

H.V. Jagadish	University of Michigan, USA
Sushil Jajodia	George Mason University, USA
Q. Jin	Waseda University, Japan
A. Kumar	Pennsylvania State University, USA
A. Mondal	Xerox Research, Bangaloru, India
K. Myszkowski	Max-Planck-Institut für Informatik, Germany
Alexander Pasko	Bournemouth University, UK
L. Pichl	International Christian University, Tokyo, Japan
P.K. Reddy	International Institute of Information Technology, Hyderabad, India
C. Shahabi	University of Southern California, USA
M. Sifer	Sydney University, Australia
F. Wang	Microsoft Research, USA

Sponsoring Institution

Center for Strategy of International Programs, University of Aizu,
Aizu-Wakamatsu City, Fukushima, Japan.

Table of Contents

Networked Information Resources I

Business Data Analytics in Astronomy and Sciences

Networked Information Resources II

The Big Data Landscape: Hurdles and Opportunities

Divyakant Agrawal[1,2] and Sanjay Chawla[1,3]

[1] Qatar Computing Research Institute, Qatar
[2] University of California Santa Barbara, USA
[3] University of Sydney, Australia

Abstract. Big Data provides an opportunity to interrogate some of the deepest scientific mysteries, e.g., how the brain works and develop new technologies, like driverless cars which, till very recently, were more in the realm of science fiction than reality. However Big Data as an entity in its own right creates several computational and statistical challenges in algorithm, systems and machine learning design that need to be addressed. In this paper we survey the Big Data landscape and map out the hurdles that must be overcome and opportunities that can be exploited in this paradigm shifting phenomenon.

1 Introduction

Big data has emerged as one of the most promising technology paradigm in the past few years. Availability of large data arises in numerous application contexts: trillions of words in English and other languages, hundreds of billions of text documents, a large number of translations of documents in one language to other languages, billions of images and videos along with textual annotations and summaries, thousands of hours of speech recordings, trillions of log records capturing human activity, and the list goes on. During the past decade, careful processing and analysis of different types of data has had transformative effect. Many applications that were buried in the pages of science fiction have become a reality, e.g., driverless cars, language agnostic conversation, automated image understanding, and most recently deep learning [6] to simulate a human brain.

In the technology context, Big Data has resulted in significant research and development challenges. From a systems perspective, scalable storage, retrieval, processing, analysis, and management of data poses the biggest challenge. From an application perspective, leveraging large amounts of data to develop models of physical reality becomes a complex problem. The interesting dichotomy is that the bigness of data in the system context makes some of the known data processing solutions that were "acceptable" to "not acceptable." For example, standard algorithms for carrying out join processing may have to be revisited in the Big Data context. In contrast, the bigness of data allows many applications to move from being "not possible" to "possible." For example real time, automated, high quality and robust language translation seems entirely feasible. Thus, new

W. Chu et al. (Eds.): DNIS 2015, LNCS 8999, pp. 1–11, 2015.

approaches are warranted to develop scalable technologies for processing and managing Big Data [7]. In the same vein, designing and developing robust models for learning from big data remains a significant research challenge.

In this paper, we explore the Big Data problem both from the system perspective as well as from the application perspective. In the popular press, "bigness" or the size of data is touted as a desirable property. Our goal is to clearly comprehend the underlying complexity that must be overcome with the increasing size of data and clearly delineate the hurdles and opportunities in this fast developing space.

The rest of the paper is structured as follows. In Section 2 we use the record de-duplication application to delve into some of the intrinsic computational, systems and statistical challenges when operating in a Big Data environment. In Section 3 we use the classical clustering problem to highlight how simple machine learning tasks can result in complex computational problems and the resulting trade-offs in learning in a Big Data environment. We briefly review the dictionary learning problem in Section 4 and its connects with deep and representational learning. We conclude in Section 5 with a discussion and directions for future work.

2   The Big Data Problem

To understand the "Big Data" problem consider a table $S(r, c)$ whose rows (r) and columns (c) can grow infinitely. Assume the growth rate of the table (in terms of r and c) outstrips the corresponding increase in unit computational processing power and storage capacity. We refer to this setting as the Big Data Operating Paradigm (BDOP).

A task $\mathbb{T}$ on table S is an operation which maps S onto another entity E, where E can be another table or the state of a mathematical model specified as part of the task $\mathbb{T}$. The Big Data problem is to understand the feasibility of carrying out $\mathbb{T}$ both in terms of computational tractability and statistical effectiveness in BDOP as S grows.

Example: Let $S(r, c)$ be a merged table of customer records from two databases. Let $\mathbb{T}$ be the task of record de-duplication, i.e., identify records in S which belong to the same customer. The computational challenge arises because all pairs of records $(O(|r|^2))$ in S have to be compared. The statistical hardness comes into fore because of the high dimensionality of the problem as the number of columns $(|c|)$ increases.

2.1   Distributed Computational Complexity

For concreteness consider the record de-duplication task [3] which can be specified as

$$\mathbb{T} : S \times S \to_d \{0, 1\}$$

Here, d is a "dis-similarity" function between two records which is either specified by a domain expert or learnt separately from the data. As noted earlier, the computational complexity of $\mathbb{T}$ is $O(|r|^2)$.

In a BDOP setting it becomes necessary that S is stored in a distributed environment and the task $\mathbb{T}$ carried out in parallel. However, if we assume that in a given time window, data grows by an order of magnitude but unit processing speed and storage capacity are constant, then for the $O(|r|^2)$ de-duplication task $\mathbb{T}$, the number of computation nodes required to maintain the same response time grows at least quadratically! This might appear as a paradox as the resources required to maintain the same quality of service is an order of magnitude greater than the corresponding increase in data size. However, in practical real data settings, even if $|S \times S|$ grows quadratically, the computational complexity of $\mathbb{T}$ is governed by $|S \bowtie S|$ which grows at the rate $O(\alpha.|r|)$, where α is typically a small fraction. Thus the task $\mathbb{T}$ becomes feasible as the corresponding bi-partite graph is sparse. However, while α maybe small, the distribution of the degree of the bi-partite graph is highly non-uniform (often Zipfian). Thus the processing time in a distributed environment is lower bounded by the size of the largest partition which can be large. Designing the appropriate trade-off between data partitioning and task parallelism in a cloud environment becomes a major design and research challenge.

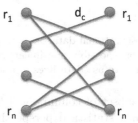

Fig. 1. From a computational and systems perspective, a Big Data task reduces to processing a table as a bi-partite graph in a distributed setting, where the edge similarity is derived as a function of the columns of a table. In a BDOP setting, computational is feasible only because the number of edges grow at constant rate as the number of rows increase, i.e., the bi-partitite graph is sparse.

2.2   Statistical Effectiveness

From a statistical perspective, we focus on the columns of S. For the de-duplication task the objective is to *infer* a function $f : \{C\} \rightarrow 2^{\{C\}}$ on the columns such that for any pair of records r_1, r_2

$$\pi_{f(C)}(r_1) = \pi_{f(C)}(r_2) \rightarrow r_1 \equiv r_2$$

We can interpret $f(C)$ as a subset of columns of S. Now, given the statistical nature of data, every subset of columns has a small probability p of being selected by the inference function f. However as $|c|$ increases then the probability that an

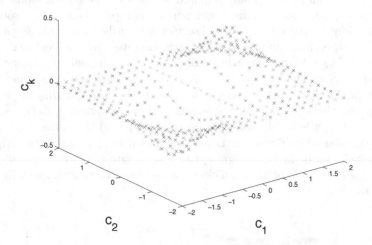

Fig. 2. Observational data tends to be highly redundant, i.e., there is a large degree of correlation (possibly non-linear) between the different columns of table S. This is often referred to as "data lives in a low-dimensional manifold."

arbitrary subset of columns will be selected by f is given $1 - (1 - p)^{2^{|c|}} \to 1$. Thus from a statistical perspective, Big Data settings can lead to situations where the danger of inferring spurious relationships becomes highly likely. However, in practice, observational data (i.e., data collected serendipitously and not as part of an experimental design), tends to be highly redundant. High redundancy implies that the number of degrees of freedom which govern data generation is small. This is often referred to as "observational data lives in a low-dimensional manifold." The manifold structure of the data explains the widespread use of dimensionality reduction techniques like PCA and NMF in machine learning and data mining.

3   Machine Learning

Machine Learning tasks are often formulated as optimization problems. Even the simplest of tasks can result in hard and intractable optimization formulations. In BDOP, the constraint on the solution is often determined by a "time budget," i.e., obtain the best possible approximate solution of the optimization problem within time T. We highlight the trade-off between large data and the quality of the optimization solution using two examples: clustering and dictionary learning. The latter in intimately tied to "deep learning."

3.1 Clustering

Consider a set of one dimensional data points: $D = \{x_1, \ldots, x_n\}$. For example, D could record the height of a group of n people. How do we summarize D? One obvious answer is the mean (average) of D, but lets cast the summarization task as an optimization problem. Thus, the objective is to find a y which minimizes the following objective function:

$$\min_{y \in \mathbb{R}} \sum_{i=1}^{n} (x_i - y)^2$$

If we denote $F(y)$ as $\sum_{i=1}^{n}(x_i - y)^2$, then a necessary condition to obtain y is to set $\frac{dF}{dy} = 0$ and solve for y to obtain:

$$y^* = \frac{1}{n} \sum_{i=1}^{n} x_i$$

Thus, as expected, the "optimal" summarization of D is to use average or mean of the data set. Now suppose we would like to summarize D with two data points, y_1 and y_2. The motivation for this task is to possibly obtain a representative male and female height in the group. Note, the data set D is not labeled, i.e.,we are not given which data point records a male or female height. In this situation, what might be an appropriate objective function? The first instinct is perhaps to set up the optimization problem where the aim is to find y_1 and y_2 which minimizes

$$G(y_1, y_2) = \sum_{i=1}^{n} \sum_{j=1}^{2} (x_i - y_j)^2$$

However, an examination of G reveals that it may not be an appropriate objective function. For example, we do not want to take the difference between every pair of x_i and y_j but *only* between x_i and its most *representative* y_j, which of course is not known. Thus a more suitable objective is to minimize:

$$H(y_1, y_2) = \sum_{i=1}^{n} \min_{j \in \{1,2\}} (x_i - y_j)^2 \tag{1}$$

The appearance of the min inside the summation, makes the objective non-convex and is a typical optimization pattern in many machine learning problem formulations. In Figure 3, the objective functions G and H are plotted which clearly show that H (the relevant objective function) is non-convex.

3.2 Clustering in BDOP

Having specified a clustering objective function in Equation 1, the question remains how to solve the resulting optimization problem and understand the impact of Big Data on the solution. An important observation is that the clustering

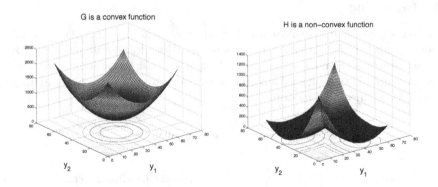

Fig. 3. The figure on the left is the shape of the objective function $G(y_1, y_2)$. G is convex but not appropriate for the clustering task. The figure on the right is the shape of $H(y_1, y_2)$ which is more suitable for clustering but non-convex.

problem is inherently imprecise. Even in the case where each data point represents a person's height and the goal is to obtain separate clusters for males and females, there are bound to be misclassified data points, i.e., males will be assigned to the female cluster and vice-versa. The availability of large amounts of data will not mitigate the imprecision and thus a case can be made to design an algorithm where the ability to trade-off between exactness of the solution and the time to obtain the solution is transparent [2]. For example, instead of optimizing $H(y_1, y_2)$, perhaps a more reasonable objective is to optimize

$$\mathbb{E}_x[H(y_1, y_2, x)] \tag{2}$$

where the expecation is over different samples of x from the same underlying (but unknown) distribution $P(x)$. Bottou et. al. [2] have precisely investigated the decomposition of Equation 2 in a general case which we customize for the case of clustering.

3.3   Error Decomposition for Big Data Learning

Notice that the objective function $H(y_1, y_2, x)$ is highly specialized as we have used $(x_i - y_j)^2$ term inside the summation. In a high-dimensional setting, this is equivalent to enforcing the resulting clusters to be spherical. Thus if the original clusters were elliptical then the choice of the objective function already results in an error which is *independent of the size of the data* ! This is known as the *approximation error*. Now because our objective is to minimize $\mathbb{E}_x[H(y_1, y_2, x)]$ but we only have access to finite number of samples from $P(x)$, the solution of any algorithm will result in value H_n such that $H_n \leq \mathbb{E}_x[H(y_1, y_2, x)] + \rho$. The discrepancy (bounded by ρ) is known as the *estimation error*. In a Big Data environment, where a given time budget may force us to stop the optimization task before all the samples are processed will lead to an objective value of $\hat{H}_n$, which is an approximation of H_n. This is known as the *computation error*.

3.4   Stochastic Gradient Descent

A simple but extremely versatile optimization algorithm where the trade-off between the estimation and computation error can be controlled is the Stochastic Gradient Descent Algorithm (SGDA). Recall that in order to optimize

$$\min_y f(x, y)$$

the gradient descent algorithm iterates on the following step till an approximate fixed point is obtained

$$y_{t+1} = y_t - \gamma \nabla f(y_t, x).$$

Now since in many machine learning formulations, $f(x, y)$ is of the form $f(x, y) = \sum_i^n g(x_i, y)$, the cost of the gradient step is $O(n)$. However, in stochastic gradient a random sample x_r from the training set is selected in each iteration to approximate the gradient

$$\hat{\nabla} f(y) \approx \nabla g(x_r, y)$$

The above approximation reduces the computation cost of the gradient computation from $O(n)$ to $O(1)$ and it has been shown that as $n \to \infty, y_{t+1} = y_t - \gamma \nabla g(x_r, y)$ approaches the optimal solution of $\min_y \mathbb{E}_x[f(x, y)]$.

The SGDA algorithm can easily be adapted to solve the clustering problem with objective function $H(y_1, y_2)$ as given in Equation 1. Here are the steps [1]:

1. Initialize y_1 and y_2 and set $n_1, n_2 = 0$.
2. Randomly permute D to obtain $\{x_{i_1}, \ldots x_{i_n}\}$
3. **For** $j = 1 : n$

   (a) Choose y_k closest to x_{i_j}:

   $$k^* = \arg\min {}_k (x_{i_j} - y_k)^2$$

   (b) Increment the count associated with y_{k^*}

   $$n_{k^*} \leftarrow n_{k^*} + 1$$

   (c) Update y_{k^*} (the gradient descent step):

   $$y_{k^*} \leftarrow y_{k^*} + \frac{1}{n_{k^*}}(x_{i_j} - y_{k^*})$$

4. **End For**

Notice that SGDA is especially suitable in a Big Data setting compared to the traditional batch k-means algorithm as it is online and can start emitting results incrementally.

4   Dictionary Learning

Some of the most promising applications of Big Data include language translation, speech recognition, image search and cross-modality querying, i.e., given
an image automatically generate a textual description of the image and given
a piece of text find the most appropriate image. Deep learning (aka neural networks) have been re-emerged as the algorithm of choice for these applications.
However, the fundamental reason for success in these applications is because of
the ability for algorithms to *learn* the appropriate *representation* in the presence
of Big Data. The learning of an appropriate representation is often called the
Dictionary Learning or sparse coding problem. We briefly describe the dictionary
learning problem and note its similarity with the clustering problem described
above. While we will not provide the algorithm, it should become clear that
SGDA can be used to solve the dictionary learning problem.

We again start with a data set $D = \{\mathbf{x_1}, \mathbf{x_2}, \ldots, \mathbf{x_n}\}$, where each $\mathbf{x_i} \in \mathbb{R}^m$.
For example the set D may be a collection of images and $\mathbf{x_i}$ is a vector of pixel
values. In signal and image processing, data is often represented as a linear
combination of *pre-defined* basis functions using Fourier or Wavelet transforms.
In dictionary learning (also called sparse coding), the objective is to find a set
of data-dependent basis functions $\mathbf{E} \in \mathbb{R}^{m \times k}$, and a set of $k-$dimensional sparse
vectors $\{\alpha_\mathbf{i}\}_{i=1}^n$,such that

$$\mathbf{x_i} \approx \mathbb{E}\alpha_\mathbf{i} \ \forall i = 1, \ldots, n$$

The columns of $\mathbb{E}$ are the basis function and α's are the weights. Note that the
columns of E are not restricted to be orthogonal and this provides a degree of
flexibility that is not available in the case of Fourier or Wavelet transform or an
SVD decomposition.

In order to infer both $\mathbb{E}$ and α, we can setup an objective function which has
to be minimized

$$g_n(\mathbb{E}, \mathbf{x}) = \sum_{i=1}^n \min_{\alpha_\mathbf{i}} \|\mathbf{x}_i - \mathbb{E}\alpha_\mathbf{i}\|_2^2 + \lambda\|\alpha_i\|_1 \tag{3}$$

Notice the similarity between the clustering objective in Equation 1 and the
dictionary learning objective in Equation 3. Both objectives have a min inside the
sum function and in the case of dictionary learning there is a coupling (product)
between the two unknowns ($\mathbb{E}$ and α) which makes the objective non-convex.
More details about dictionary and represenational learning can be found in [9,5].

4.1   Distributed Stochastic Gradient Descent

Many machine learning problems are formulated as optimization problems with
a very specific form. For example a typical optimization pattern will be of the
form

$$\sum_{i=1}^n \ell(x_i, \mathbf{w}) + \Omega(\mathbf{w}) \tag{4}$$

Here, $\ell(x, \mathbf{w})$ is the loss function which accounts for the mismatch between the data and the model. The regularization term $\Omega(\mathbf{w})$ is often used to prevent overfitting of the model to the data. An important observation is that the loss function is applied pointwise to each data point and this can be used to design efficient distributed implementations.

A common distributed computation design pattern for machine learning is the *parameter server* pattern which consists of the following components: (i) data is partitioned (horizontally) and distributed across computational nodes, (ii) a parameter server node maintains the global state of the variable $\mathbf{w}$, (iii) each computation node i applies (stochastic) gradient descent to its data partition and computes a gradient $\nabla_i(w)$, (iv) the parameter server periodically polls the compute nodes and pulls ∇_i from each node. It then carries out a global 'sync' operation and pushes the updated $\mathbf{w}$ vector to the nodes [4,8].

The nature of machine learning problems tasks i that they can afford to tolerate a level of imprecision which is not available in other application domains (e.g., airplane engine simulation). Together with specific optimization pattern that emerges in many machine learning provides an opportunity for creating "near embarrassingly parallel" implementations.

5   Discussion and Conclusion

The availability of Big Data across many application domains has led to the promise of designing new applications which hitherto were considered out of reach. For example, Big Data has been instrumental in designing algorithms for real time language translation, high quality speech recognition and image and video search. Large amount of FMRI and MEG brain data collected while people are carrying out routine tasks holds the promise of understanding the working of the brains.

To fully utilize the promise of Big Data several hurdles in the computational, systems and algorithmic aspects of data processing have to be overcome. In the Big Data Operating Paradigm (BDOP) the growth rate of data is higher than the corresponding increase in computational processing speed and storage capacity. This necessarily leads to an environment where data has to be processed in a distributed manner. Since many algorithm for data processing and machine learning are super-linear, increase in data size results in a much higher increase in infrastructure resources - if quality of service constraints have to be met. From a computational perspective many data analytic tasks can be abstracted as the processing of a bipartite graph where the nodes are rows of a table and the edges are determined by a similarity function based on the columns. If the edges of the graph were uniformly distributed then processing in BDOP would be impossible. However bipartite graphs of real data tends to be highly sparse and skewed. Sparsity provides an opportunity to process data efficiently in BDOP but skewness results in a lower bound on the computation. The interplay between sparsity and skewness is a major systems challenge that is currently addressed on a case by case basis.

From a statistical perspective, availability of Big Data does not necessarily result in better quality or stable results. In fact the danger of deriving spurious relationships are greater in a high-dimensional than a low-dimensional setting. For real data it has been observed that even though the ambient dimension of the data may be high, its intrinsic dimension is low. This is often referred to as "data living in a low dimensional manifold." The low intrinsic dimensionality explains the widespread use of dimensionality reduction techniques like SVD and NMF. Most dimensionality reduction techniques have high computational complexity (often $O(n^3)$). Sometimes the use of random projection in conjunction with dimensionality reduction can lead to improved efficiency without substantial loss in accuracy as machine learning task output can afford a degree of imprecision which is greater than in many other domains (like aircraft engine design or exactness of OLTP query result).

Machine Learning tasks are often cast as optimization problems. Even the simplest of tasks, like data summarization, can result in complex optimization formulations. The intrinsic coupling between the approximation, estimation and computational error while solving the optimization task has several implications including the ability to use simple first order algorithms like stochastic gradient descent which trade-off between estimation and computational error in a transparent manner. For example, the typical cost of solving the k-means algorithm on a data set of size n is $O(nI)$ where I is the number of iterations. In BDOP the standard k-means algorithms is near impossible to use as it requires multiple passes over the data. However, SGDA can start producing reasonably accurate clusters within one pass and time budget can be used to settle on the quality of the result. Furthermore the form of a typical optimization objective function and tolerance of a certain amount of imprecision makes machine learning tasks amenable to highly efficient distributed and parallel implementations.

In the last five years there has been a resurgence of interest in deep learning. While deep learning is synonymous with neural networks, the key insight is to *infer* a representation of raw data specific for the task at hand. This is often referred to as dictionary learning or sparse coding. The dictionary learning problem can be cast as optimization problem where the objective has a similar form like the clustering problem. SGDA algorithms have been successfully employed for dictionary learning which attest to their versatility.

A grand challenge in the Big Data landscape continues to be a lack of a system which has the robustness and scalability of a traditional relational database management system while offering the expressive power to model a large and customizable class of machine learning problems.

References

1. Bottou, L., Bengio, Y.: Convergence properties of the k-means algorithms. In: Advances in Neural Information Processing Systems Conference, NIPS, Denver, Colorado, USA, pp. 585–592 (1994)

2. Bottou, L., Bousquet, O.: The tradeoffs of large scale learning. In: Proceedings of the Twenty-First Annual Conference on Neural Information Processing Systems, Advances in Neural Information Processing Systems 2007, Vancouver, British Columbia, Canada, December 3-6, pp. 161–168 (2007)
3. Christen, P.: A survey of indexing techniques for scalable record linkage and deduplication. IEEE Trans. Knowl. Data Eng. 24(9), 1537–1555 (2012)
4. Dean, J., Corrado, G., Monga, R., Chen, K., Devin, M., Mao, M.: Large scale distributed deep networks. In: Pereira, F., Burges, C., Bottou, L., Weinberger, K. (eds.) Advances in Neural Information Processing Systems, vol. 25, pp. 1223–1231 (2012)
5. Goodfellow, I.J., et al.: Challenges in representation learning: A report on three machine learning contests. In: Lee, M., Hirose, A., Hou, Z.-G., Kil, R.M. (eds.) ICONIP 2013, Part III. LNCS, vol. 8228, pp. 117–124. Springer, Heidelberg (2013)
6. Hinton, G.E.: Deep belief nets. In: Encyclopedia of Machine Learning, pp. 267–269 (2010)
7. Kraska, T., Talwalkar, A., Duchi, J.C., Griffith, R., Franklin, M.J., Jordan, M.I.: Mlbase: A distributed machine-learning system. In: Proceedings of the Sixth Biennial Conference on Innovative Data Systems Research, CIDR 2013, Asilomar, CA, USA, January 6-9 (2013)
8. Li, M., Andersen, D.G., Park, J.W., Smola, A.J., Ahmed, A., Josifovski, V., Long, J., Shekita, E.J., Su, B.-Y.: Scaling distributed machine learning with the parameter server. In: 11th USENIX Symposium on Operating Systems Design and Implementation, OSDI 2014, Broomfield, CO, USA, October 6-8, pp. 583–598 (2014)
9. Mairal, J., Bach, F., Ponce, J., Sapiro, G.: Online learning for matrix factorization and sparse coding. Journal of Machine Learning Research 11, 19–60 (2010)

Discovering Chronic-Frequent Patterns in Transactional Databases

R. Uday Kiran[1,2] and Masaru Kitsuregawa[1,3]

[1] Institute of Industrial Science, The University of Tokyo, Tokyo, Japan
[2] National Institute of Information and Communication Technology, Tokyo, Japan
[3] National Institute of Informatics, Tokyo, Japan
{uday_rage,kitsure}@tkl.iis.u-tokyo.ac.jp

Abstract. This paper investigates the partial periodic behavior of the frequent patterns in a transactional database, and introduces a new class of user-interest-based patterns known as chronic-frequent patterns. Informally, a frequent pattern is said to be **chronic** if it has sufficient number of cyclic repetitions in a database. The proposed patterns can provide useful information to the users in many real-life applications. An example is finding chronic diseases in a medical database. The chronic-frequent patterns satisfy the anti-monotonic property. This property makes the pattern mining practicable in real-world applications. The existing pattern growth techniques that are meant to discover frequent patterns cannot be used for finding the chronic-frequent patterns. The reason is that the tree structure employed by these techniques' capture only the frequency and disregards the periodic behavior of the patterns. We introduce another pattern-growth algorithm which employs an alternative tree structure, called Chronic-Frequent pattern tree (CFP-tree), to capture both frequency and periodic behavior of the patterns. Experimental results show that the proposed patterns can provide useful information and our algorithm is efficient.

Keywords: Data mining, knowledge discovery in databases, frequent patterns and periodic patterns.

1 Introduction

A time series is a collection of events obtained from sequential measurements overtime. Periodic patterns are an important class of regularities that exist in a time series. Periodic pattern mining involves discovering all those patterns that have exhibited either complete or partial cyclic repetitions in a time series. Periodic pattern mining has several real-world applications including prediction, forecasting and detection of unusual activity. A classic application is market-basket analysis. It analyzes how regularly items are being purchased by the customers. For example, if the customers are purchasing '*Bread*' and '*Jam*' together at every hour of a day, then the set {*Bread, Jam*} represents a periodic pattern.

W. Chu et al. (Eds.): DNIS 2015, LNCS 8999, pp. 12–26, 2015.

The problem of finding periodic patterns has been widely studied in [1–6]. The basic model used in all of these studies, however, remains the same and is as follows:

1. Split the given time series into distinct subsets (or periodic-segments) of a fixed length.
2. Discover all periodic patterns that satisfy the user-defined *minimum support* (*minSup*). The *minSup* controls the minimum number of periodic-segments in which a pattern must appear.

Example 1. Given the time series $TS = a\{bc\}baebace$ and the user-defined *period* as 3, TS is divided into three periodic-segments: $TS_1 = a\{bc\}b, TS_2 = aeb$ and $TS_3 = ace$. Let $\{a \star b\}$ be a pattern, where '$\star$' denotes a wild character that can represent any single set of events. This pattern appears in the *periods* of TS_1 and TS_2. Therefore, its *support* count is 2. If the user-defined *minSup* count is 2, then $\{a \star b\}$ represents a periodic pattern. In the above time series, we have applied braces only for the events having more than one item for brevity. An event represents a set of items (or an itemset) having some occurrence order.

The popular adoption and successful industrial application of this basic model suffers from the following issues.

- The basic model considers time series as a symbolic sequence. As a result, this model fails to consider the actual temporal information of the events within a sequence.
- This model suffers from the sparsity problem. That is, most of the discovered patterns contain many wild characters with a very few number of events. For example, $a \star \star \star \star \star \star \star \star \star \star \star \star \star \star bc \star \star \star \star \star \star \star \star \star \star a \star \star \star \star \star \star \star \star b$. This problem makes the discovered patterns impracticable in applications.
- The periodic patterns satisfy the *anti-monotonic property* [7]. That is, all non-empty subsets of a periodic pattern are also periodic patterns. However, this property is insufficient to make the pattern mining practical or computationally inexpensive in the case of time series. The reason is number of frequent i-patterns shrink slowly (when $i > 1$) as i increases in a time series. The slow speed of decrease in the number of frequent i-patterns is due to the strong correlation between frequencies of patterns and their sub-patterns [2].

To confront these issues, researchers have introduced periodic-frequent pattern mining which involves discovering all those frequent patterns that have exhibited complete cyclic repetitions in a temporally ordered transactional database [8–11]. As the real-world is generally imperfect, we have observed that the existing periodic-frequent pattern mining algorithms cannot discover those interesting frequent patterns that have exhibited partial cyclic repetitions in a database.

With this motivation, this paper investigates the partial periodic behavior of the frequent patterns in a transactional database, and introduce a class of user-interest-based patterns known as **chronic-frequent patterns**. Informally, a frequent pattern is said to be **chronic-frequent** if it has sufficient number of cyclic repetitions in a database. A novel measure, called *periodic-recurrence*, has been introduced in this paper. This measure assess the periodic interestingness of a frequent pattern with respect to the number of cyclic repetitions in the entire database. The patterns discovered with this measure satisfy the anti-monotonic property. That is, all non-empty subsets of a chronic-frequent pattern are also chronic-frequent. This property makes the chronic-frequent pattern mining practicable in real-life applications. The existing pattern-growth techniques that are meant to discover frequent patterns in a transactional database [12] cannot be used for finding the chronic-frequent patterns. It is because the tree structure used by these techniques' capture only the frequency and disregard the periodic behavior of the patterns. In this paper, we have introduced another tree structure, called Chronic-Frequent Pattern Tree (CFP-tree), to capture both frequency and periodic behavior of the patterns. A pattern-growth algorithm, called Chronic-Frequent pattern-growth (CFP-growth), has been proposed to discover the patterns from CFP-tree. Experimental results show that CFP-growth is runtime efficient and scalable as well.

The rest of the paper is organized as follows. Section 2 describes the related work on periodic pattern mining. Section 3 introduces our model of chronic-frequent patterns. Section 4 describes the working of CFP-growth algorithm. The experimental evaluation of CFP-growth has been presented in Section 5. Finally, Section 6 concludes the paper with future research directions.

2   Related Work

Finding periodic patterns has been widely investigated in various domains as temporal patterns [13] and cyclic association rules [14]. These approaches discover all those patterns which are exhibiting complete cyclic repetitions in a time series data. Since the real-world is imperfect, Han et al. [1] have introduced a model to find periodic patterns which are exhibiting either complete or partial cyclic repetitions in a time series. Later, they have proposed the *max-subpattern hit set property* to reduce the computational cost of finding the periodic patterns [2]. Berberidis et al. [4] and Cao et al. [5] have tried to address an open problem of specifying the *period* using autocorrelation and other methods. Yang et al. [3] have used **information gain** to discover periodic patterns involving both frequent and rare items. All of these approaches consider time series as a symbolic sequence, and therefore, do not consider the actual temporal information of the events within a series.

Tanbeer et al. [8] have represented each event in time series as a pair constituting of an itemset and its timestamp. Next, they have modeled time series as a temporally ordered transactional database, and investigated the **full periodic behavior** of the frequent patterns to discover a class of user-interest-based patterns known as *periodic-frequent patterns*. The approach of representing time series as a transactional database has the following advantages:

- Symbolic sequences do not consider the temporal information of the events within a time series. On contrary, the same information is considered in temporally ordered transactional databases.
- The *anti-monotonic property* can not effectively reduce the search space in symbolic sequences [2]. However, the same property reduces the search space effectively in transactional databases.
- Fast algorithms, such as pattern-growth technique, can be employed to discover the patterns efficiently.

Recently, Chen et al. [15] have shown that by representing a symbolic sequence as a transactional database, one can employ a pattern-growth technique to outperform the *max-subpattern hit set algorithm* [2]. Thus, many researchers are extending Tanbeer's work to address the *rare item problem* [9, 10] and top$-k$ [11] periodic pattern mining. All of the above approaches try to discover those frequent patterns that are exhibiting complete cyclic repetitions in the entire database. On the contrary, our model focuses on finding the frequent patterns that are **exhibiting either complete or partial cyclic repetitions in a database**.

In [16, 17], we have introduced a measure known as *periodic-ratio* to assess the partial periodic behavior of a frequent pattern. Unfortunately, finding the patterns with this measure is a computationally expensive process because the discovered patterns do not satisfy the anti-monotonic property. In this paper, we have introduced an alternative interestingness measure which not only assess the partial periodic behavior of a frequent pattern, but also ensures that the discovered patterns satisfy the anti-monotonic property.

Overall, the proposed model of finding chronic-frequent patterns in a transactional database is novel and distinct from the existing models.

3   Proposed Model

Let $I = \{i_1, i_2, \cdots, i_n\}$ be the set of items. Let $X \subseteq I$ be a **pattern**. A pattern containing k number of items is called a k-**pattern**. A **transaction**, $tr = (tid, Y)$, is a tuple, where *tid* represents the transaction-identifier (or a **timestamp**) and Y is a pattern. A transactional database TDB over I is a set of transactions $T = \{t_1, t_2, \cdots, t_m\}$, $m = |TDB|$, where $|TDB|$ represents the size of TDB in total number of transactions. For a transaction $tr = (tid, Y)$, such that $X \subseteq Y$, it is said that X occurs in tr and such transaction-identifier is denoted as tid^X. Let $TID^X = \{tid_j^X, \cdots, tid_k^X\}$, $j, k \in [1, m]$ and $j \leq k$, be the set of all transaction-identifiers at which X has appeared in TDB. The size of TID^X is defined as the *support* of X, and denoted as $S(X)$. That is, $S(X) = |TID^X|$. The pattern X is said to be **frequent** if $S(X) \geq minSup$, where $minSup$ is the user-defined minimum support threshold.

Example 2. Consider the transactional database shown in Table 1. It contains 10 transactions. The *tid* of each transaction represents its sequential occurrence order with respect to a particular timestamp. The set of items, $I =$

$\{a, b, c, d, e, f, g, h\}$. The set of items, 'a' and 'b', i.e., '$\{a, b\}$' is known as an itemset (or a pattern). This pattern contains two items. Therefore, it is a 2-pattern. For brevity, we refer this pattern as 'ab'. The pattern 'ab' appears in the transactions having $tids$ 1, 3, 5, 8 and 10. Therefore, $TID^{ab} = \{1, 3, 5, 8, 10\}$. The support of 'ab' is the size of TID^{ab}. Therefore, $S(ab) = |TID^{ab}| = |\{1, 3, 5, 8, 10\}| = 5$. If the user-defined $minSup = 4$, then 'ab' is a frequent pattern as $S(ab) \geq minSup$.

Table 1. Transactional database

TID	Items	TID	Items	TID	Items	TID	Items	TID	Items
1	a, b, h	3	a, b, g	5	a, b, c, d	7	c, d, h	9	c, d, g
2	e, f	4	e, f, h	6	e, f, g	8	a, b, c, d	10	a, b, c, d

Definition 1. *(A period of pattern X.) Let tid_p^X and tid_q^X, $p, q \in [1, m]$ and $p < q$, be the two consecutive transaction-ids where X has appeared in TDB. The number of transactions (or the time difference) between tid_p^X and tid_q^X can be defined as a **period** of X, say p_i^X. That is, $p_i^X = tid_q^X - tid_p^X$.*

Example 3. Continuing with Example 2, the pattern 'ab' has consecutively appeared in the $tids$ of 1 and 3. Therefore, a period of 'ab', i.e., $p_1^{ab} = 2 (= 3 - 1)$. Similarly, the other periods of 'ab' are as follows: $p_2^{ab} = 2 (= 5 - 3)$, $p_3^{ab} = 3 (= 8 - 5)$ and $p_4^{ab} = 2 (= 10 - 8)$.

Definition 2. *(An interesting period of pattern X.) Let $P^X = \{p_1^X, p_2^X, \cdots, p_k^X\}$, $k = S(X) - 1$, be the complete set of all periods of X in TDB. A $p_j^X \in P^X$ is said to be interesting iff $p_j^X \leq maxPrd$, where $maxPrd$ refers to the user-defined maximum period threshold. This definition captures the periodic occurrences of a pattern in the database.*

Example 4. The complete set of periods for 'ab', i.e., $P^{ab} = \{2, 2, 3, 2\}$. If the user-defined $maxPrd = 2$, then p_1^{ab} is an interesting period because $p_1^{ab} \leq maxPrd$. Similarly, p_2^{ab} and p_4^{ab} are interesting periods, however, p_3^{ab} is not an interesting period as $p_3^{ab} \nleq maxPrd$.

Definition 3. *(The periodic-recurrence of pattern X.) Let $IP^X \subseteq P^X$ be the set of periods such that $\forall p_j^X \in IP^X$, $p_j^X \leq maxPrd$. The size of IP^X gives the periodic-recurrence of X, say $PR(X)$. That is, $PR(X) = |IP^X|$.*

Example 5. The complete set of all interesting periods of 'ab', i.e., $IP^{ab} = \{p_1^{ab}, p_2^{ab}, p_4^{ab}\}$. Therefore, the *periodic-recurrence* of 'ab', i.e., $PR(ab) = |IP^{ab}| = 3$.

The above definition measures the number of periodic occurrences of a pattern X in TDB. Now, we define chronic-frequent patterns using the *support* and *periodic-recurrence* measures.

Definition 4. *(The chronic-frequent pattern X.) The frequent pattern X is said to be **chronic-frequent** if its periodic-recurrence is no less than the user-defined minimum periodic-recurrence threshold (minPR). That is, X is a chronic-frequent pattern if $S(X) \geq minSup$ and $PR(X) \geq minPR$.*

Example 6. If the user-defined $minPR = 3$, then the frequent pattern 'ab' is a chronic-frequent pattern as $PR(ab) \geq minPR$.

The *support* and a *period* of a pattern can be normalized to the scale of [0%, 100%] by expressing them in the percentage of $|TDB|$. Similarly, the *periodic-recurrence* of pattern X can also be normalized to the same scale by expressing it in percentage of $|TDB| - 1$, where $|TDB| - 1$ represents the maximum number of periods a pattern can have in a database (see Property 3). The patterns discovered with this normalization method satisfy the anti-monotonic property. The correctness of our argument is based on the Properties 1, 2 and 3 and shown in Lemma 1.

Property 1. The total number of *periods* for a pattern X, i.e., $|P^X| = S(X) - 1$.

Property 2. (Apriori property [7]) If $X \subset Y$, then $TID^X \supseteq TID^Y$.

Property 3. The maximum support a pattern X can have in a database is $|TDB|$. From Property 1, it turns out that the maximum number of *periods* a pattern X can have in a database is $|TDB| - 1$.

Lemma 1. *Let X and Y be the patterns such that $X \subset Y$. If $S(X) < minSup$ and $PR(X) < minPR$, then $S(Y) < minSup$ and $PR(Y) < minPR$.*

Proof. If $X \subset Y$, then $TID^X \supseteq TID^Y$ (see Property 2). Thus, $P^X \supseteq P^Y$, $IP^X \supseteq IP^Y$, $S(X) \geq S(Y)$ and $PR(X) \geq PR(Y)$ $\left(= \frac{|IP^X|}{|TDB|-1} \geq \frac{|IP^Y|}{|TDB|-1}\right)$. Therefore, if $S(X) < minSup$ and $PR(X) < minPR$, then $S(Y) < minSup$ and $PR(Y) < minPR$. Hence proved.

Definition 5. *(**Problem definition.**) Given a transactional database (TDB) and the user-defined minimum support (minSup), maximum period (maxPrd) and minimum periodic-recurrence (minPR) thresholds, discover the complete set of chronic-frequent patterns having support and periodic-recurrence no less than the minSup and minPR, respectively.*

In the next section, we discuss our algorithm to discover the complete set of chronic-frequent patterns from a transactional database.

4   The CFP-Growth Algorithm

The CFP-growth algorithm involves two steps: (*i*) compressing the database into a tree-structure, called CFP-tree and (*ii*) recursive mining of CFP-tree to discover the patterns. Before describing these two steps, we explain the structure of CFP-tree.

4.1   Structure of CFP-Tree

The CFP-tree includes a prefix-tree and a chronic-frequent item (or 1-pattern) list, called CFP-list. The CFP-list consists of four fields – *item name* (I), *total support* (f), *periodic-recurrence* (pr) and a pointer pointing to the first node in the prefix-tree carrying the item.

The prefix-tree in CFP-tree resembles the prefix-tree in FP-tree. However, to capture both *frequency* and *chronic* behaviour of the patterns, the nodes in CFP-tree explicitly maintains the occurrence information for each transaction by keeping an occurrence transaction-id list, called *tid*-list. **To achieve memory efficiency, only the last node of every transaction maintains the *tid*-list.** Hence, there are two types of nodes maintained in a CFP-tree: ordinary node and *tail*-node. The former is the type of nodes similar to that used in FP-tree, whereas the latter is the node that represents the last item of any sorted transaction. The structure of *tail*-node is $I[tid_1, tid_2, \cdots, tid_m]$, where I is the node's item name and tid_i, $i \in [1, m]$, (m be the total number of transactions from the *root* up to the node) is a transaction-id where item I is the last item. The conceptual structure of CFP-tree is shown in Figure 1. Like in FP-tree, each node in a CFP-tree maintains parent, child and node traversal pointers. However, irrespective of the node type, no node in a CFP-tree maintains support value in it.

Fig. 1. The conceptual structure of prefix-tree in CFP-tree. The dotted ellipse represents the ordinary node, while the other ellipse represents the tail-node of sorted transactions with tids.

To facilitate high degree of compactness, items in a CFP-tree are arranged in support-descending item order. It has been proved in [18] that such tree can provide a highly compact tree structure, and an efficient mining phase using pattern-growth technique.

One can assume that the structure of prefix-tree in CFP-tree may not be memory efficient as it explicitly maintains tids of each transaction. However, it has been argued in the literature [8] that such a tree can achieve memory efficiency by keeping transaction information only at the *tail*-nodes and avoiding the support count field at each node. Furthermore, CFP-tree avoids the *complicated combinatorial explosion problem of candidate generation* as in Apriori-like algorithms [7]. In the literature, keeping the information pertaining to transactional-identifiers in a tree can also been found in efficient frequent pattern mining [19, 20].

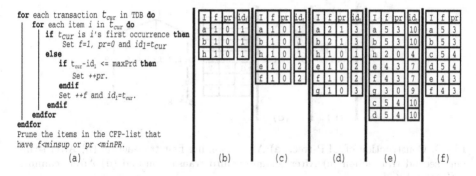

Fig. 2. Construction of CFP-list. (a) Procedure (b) After scanning first transaction (c) After scanning second transaction (d) After scanning third transaction (e) After scanning every transaction and (f) Sorted list of chronic-frequent items.

4.2 Construction of the CFP-Tree

Since chronic-frequent patterns satisfy the anti-monotonic property, chronic-frequent items (or 1-patterns) play a key role in efficient mining of these patterns. Using the CFP-list, we perform a scan on the database to discover these items. Let t_{cur} denote the *tid* of current transaction. Let id_l be a temporary array that explicitly records the *tids* of last occurring transactions of all items in the CFP-list. Figure 2(a) shows the procedure followed to discover the chronic-frequent items. We illustrate this procedure using the database shown in Table 1.

The scan on the first transaction '1 : a, b, h', with $t_{cur} = 1$, inserts the items 'a', 'b' and 'h' into the CFP-list with $f = 1$, $pr = 0$ and $id_l = 1$ (see Figure 2(b)). The scan on the second transaction '2 : e, f', with $t_{cur} = 2$, inserts the items 'e' and 'f' into the CFP-list with $f = 1$, $pr = 0$ and $id_l = 2$ (see Figure 2(c)). The scan on the third transaction '3 : a, b, g', with $t_{cur} = 3$, adds the item 'g' into the CFP-list with $f = 1$, $pr = 0$ and $id_l = 3$. Simultaneously, the 'f', 'pr' and 'id_l' values of 'a' and 'b' are updated to 2, 1 and 3, respectively. Figure 2(d) shows the CFP-list constructed after scanning the third transaction. Similar approach is followed for the remaining transactions and CFP-list is updated accordingly. Figure 2(e) shows the CFP-list constructed after scanning all transactions in the database. The items having *support* less than the *minSup* or *periodic-recurrence* less than the *minPR* are pruned from the CFP-list. The remaining items are sorted in descending order of their frequencies. Figure 2(f) shows the sorted list of chronic-frequent items in CFP-list. Let CF denote this sorted list of items.

Using the FP-tree construction technique, only the items in the CF will take part in the construction of CFP-tree. The *tree* construction starts by inserting the first transaction, '1 : a, b, h', according to CFP-list order, as shown in Figure 3(a). The tail-node '$b : 1$' carries the *tid* of the transaction. Please note that the item 'h' was not considered in the construction of CFP-tree as it is not a chronic-frequent item. Similar process is repeated for other transactions in the database. Figure 3(b), (c) and (d) respectively show the CFP-tree constructed after scanning second transaction, third transaction and entire database. For

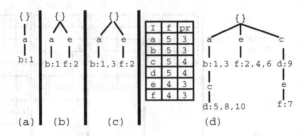

Fig. 3. Construction of CFP-tree. (a) After scanning first transaction (b) After scanning second transaction (c) After scanning third transaction and (d) After scanning entire database.

the simplicity of figures, we do not show the node traversal pointers in trees, however, they are maintained in a fashion like FP-tree does.

The CFP-tree explicitly maintains *tids* of each transaction at the nodes. As a result, one can argue that the structure of a CFP-tree may not be memory efficient. We argue that the CFP-tree achieves the memory efficiency by keeping such transaction information only at the *tail*-nodes and avoiding the support count field at the each node. It was also shown in the literature [8] such trees are memory efficient. Moreover, keeping the *tid* information in tree can also be found in the literature for efficient mining of frequent patterns [19].

4.3 Mining CFP-Tree

Even though both CFP-tree and FP-tree arrange items in support-descending order, we can not directly apply the FP-growth mining on a CFP-tree. The reason is that, CFP-tree does not maintain the support count at each node, and it handles the *tid*-lists at *tail*-nodes. Therefore, we devise an alternative pattern growth-based bottom-up mining technique that can handle the additional features of CFP-tree.

The basic operations in mining CFP-tree involves (*i*) counting length-1 chronic-frequent items, (*ii*) constructing the prefix-tree for each chronic-frequent patterns, and (*iii*) constructing the conditional tree from each prefix-tree. The CFP-list provides the length-1 chronic-frequent items. Before discussing the prefix-tree construction process we explore the following important property and lemma of a CFP-tree.

Property 4. A tail-node in a CFP-tree maintains the occurrence information for all the nodes in the path (from that tail-node to the root) at least in the transactions in its tid-list.

Lemma 2. Let $Z = \{a_1, a_2, \cdots, a_n\}$ be a path in a CFP-tree where node a_n is the tail-node carring the tid-list of the path. If the tid-list is pushed-up to the node a_{n-1}, then a_{n-1} maintains the occurrence information of the path $Z' = \{a_1, a_2, \cdots, a_{n-1}\}$ for the same set of transactions in the tid-list without any loss.

Proof. Based on the Property 4, a_n maintains the occurrence information of the path Z' at least in the transactions in its *tid*-list. Therefore, the same *tid*-list at node a_{n-1} exactly maintains the same transaction information for Z' without any lose.

Choosing the last item 'i' in the CFP-list, we construct its prefix-tree, say PT_i, with the prefix sub-paths of nodes labeled 'i' in the CFP-tree. Since 'i' is the bottom-most item in the CFP-list, each node labeled 'i' in the CFP-tree must be a *tail*-node. While constructing the PT_i, based on Property 4, we map the *tid*-list of every node of 'i' to all items in the respective path explicitly in a temporary array. It facilitates the calculation of *support* and *periodic-recurrence* for each item in the CFP-list of PT_i. Moreover, to enable the construction of the prefix-tree for the next item in the CFP-list, based on Lemma 2, the *tid*-lists are pushed-up to respective parent nodes in the original CFP-tree and in PT_i as well. All nodes of i in the CFP-tree and i's entry in the CFP-list are deleted thereafter. Figure 4 (a) shows the prefix-tree of 'f', i.e., PT_f. Figure 4(c) shows the status of the CFP-tree of Figure 3(d) after removing the bottom-most item 'f'.

The conditional tree CT_i for PT_i is constructed by removing all non-chronic-frequent items from the PT_i. If the deleted node is a tail-node, its *tid*-list is pushed-up to its parent node. Figure 4(b) shows the conditional tree for 'f', CT_f constructed from the PT_f of Figure 4(a). The contents of the temporary array for the bottom item 'j' in the CFP-list of CT_i represent the TID^{ij} (i.e., the set of all *tids* where item i and j occur together in the database). Therefore, it is rather simple calculation to compute $S(ij)$ and $PR(ij)$ from TID^{ij}. If $S(ij) \geq minSup$ and $PR(ij) \geq minPR$, then the pattern 'ij' is generated as a chronic-frequent pattern. The same process of creating prefix-tree and its corresponding conditional tree is repeated for further extensions of 'ij'. The whole process of mining for each item is repeated until CFP-list $\neq \emptyset$.

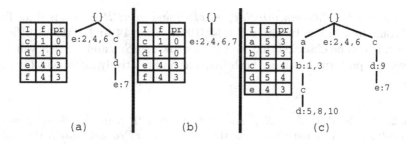

(a)   (b)   (c)

Fig. 4. Prefix-tree and conditional tree construction with CFP-tree. (a) Prefix-tree for 'f' (b) Conditional tree for 'f' and (c) CFP-tree after removing item 'f'.

5   Experimental Results

Since there is no existing approach to discover chronic-frequent patterns, we only investigate the performance of CFP-growth algorithm. In addition, we also discuss the usefulness of proposed patterns using a real-world database.

The CFP-growth algorithm was written in Java and run on Ubuntu on a 2.66 GHz machine having 4GB of memory. The databases used for our experiments are as follows:

- **T10I4D100K and T10I4D1000K Databases.** These two databases are synthetic transactional databases generated using the procedure given in [7]. The T10I4D100K dataset contains 100,000 transactions and 941 distinct items. The T10I4D1000K contains 983,155 transactions with 30,387 items.
- **Shop-14 Database.** A Czech company has provided clickstream data of seven internet shops in ECML/PKDD 2005 Discovery challenge [21]. In this paper, we have considered the click stream data of product categories visited by the users in "Shop 14" (www.shop4.cz), and created a transactional database with each transaction representing the set of web pages visited by the people at a particular *minute interval*. The transactional database contains 59,240 transactions (i.e., 41 days of page visits) and 138 product categories (or items).
- **BMS-WebView-1 Database.** This is a real-world database containing 59,602 transactions with 497 items [22].
- **Kosarak Database.** This is a very large real-world database containing 990,002 transactions with 41,270 distinct items.

The *Kosarak* and *BMS-WebView-1* databases have been downloaded from the Frequent Itemset MIning (FIMI) repository (http://fimi.ua.ac.be/data/).

5.1   Generation of Chronic-Frequent Patterns

Table 2 shows the different *minSup*, *maxPrd* and *minPR* values used for finding chronic-frequent patterns in T10I4D100K, Shop-14 and BMS-WebView-1 datasets. It can be observed that we have set low *minSup* and *minPR* values to discover the patterns involving both frequent and relatively infrequent (or rare) items.

Table 2. The user-defined *minSup*, *maxPrds* and *minPR* values in different datasets. The Greek letters α, β and γ represent the *minSup*, *maxPrd* and *minPR* thresholds, respectively.

Datasets	minSup (α)			maxPrd (β)			minPR (γ)		
	α_1	α_2	α_3	β_1	β_2	β_3	γ_1	γ_2	γ_3
T10I4D100K	0.1%	0.3%	0.5%	1%	5%	10%	0.1%	0.2%	0.3%
Shop-14	0.1%	0.3%	0.5%	1%	5%	10%	0.1%	0.2%	0.3%
BMS-WebView-1	0.1%	0.3%	0.5%	1%	5%	10%	0.1%	0.2%	0.3%

Table 3. The number of chronic-frequent patterns generated at different $minSup$, $maxPrd$ and $minPR$ threshold values

Dataset	α	γ_1			γ_2			γ_3		
		β_1	β_2	β_3	β_1	β_2	β_3	β_1	β_2	β_3
T10I4D100K	α_1	20077	26384	26511	10115	12643	12644	3768	4432	4432
	α_2	4476	4476	4476	4476	4476	4476	3768	4432	4432
	α_3	1069	1069	1069	1069	1069	1069	1069	1069	1069
Shop-14	α_1	1215	27320	30382	4268	6377	6537	2428	3000	3058
	α_2	3089	3089	3089	3089	3089	3089	2428	3000	3058
	α_3	1244	1244	1244	1244	1244	1244	1244	1244	1244
BMS-WebView-1	α_1	1410	3227	3680	572	777	796	362	431	432
	α_2	435	435	435	435	435	435	362	431	432
	α_3	201	201	201	201	201	201	201	201	201

Table 4. Runtime requirements of CFP-growth. The runtime is expressed in seconds.

Dataset	α	γ_1			γ_2			γ_3		
		β_1	β_2	β_3	β_1	β_2	β_3	β_1	β_2	β_3
T10I4D100K	α_1	207	263	268	105	126	136	37	44	44
	α_2	45	45	45	45	45	45	37	43	43
	α_3	19	19	19	19	19	19	19	19	19
Shop-14	α_1	121	220	303	42	63	65	24	30	32
	α_2	30	30	30	30	30	30	24	32	33
	α_3	14	14	14	14	14	14	14	14	14
BMS-WebView-1	α_1	103	257	287	97	189	234	182	166	156
	α_2	78	91	251	58	79	165	38	43	98
	α_3	71	92	124	55	69	83	20	34	79

Table 3 shows the number of chronic-frequent patterns generated in different datasets at various $minSup$, $maxPrd$ and $minPR$ threshold values. The following observations can be drawn from this table. (i) At a fixed $maxPrd$ and $minPR$, increase in $minSup$ has decreased the number of chronic-frequent patterns. (ii) At a fixed $minSup$ and $minPR$, increase in $maxPrd$ has increased the number of chronic-frequent patterns. It is because the occurrences of a frequent pattern which were earlier (i.e., at low $maxPrd$ threshold) considered as aperiodic have been considered as periodic with in $maxPrd$ threshold. (iii) At a fixed $minSup$ and $maxPrd$, increase in $minPR$ has decreased the number of chronic-frequent patterns. The reason is that many frequent patterns were unable to occur periodically for longer time durations in a database.

Table 4 shows the runtime taken by CFP-growth to discover chronic-frequent patterns in T10I4D100K, Shop-14 and BMS-WebView-1 datasets. The runtime involves both the construction and mining of CFP-tree. The changes on the $minSup$, $minPR$ and $maxPrd$ shows the similar effect on runtime consumption as that of the generation of chronic-frequent patterns. It can be observed that the proposed algorithm discovers the complete set of chronic-frequent patterns at a reasonable runtime even at low $minSup$ and $minPR$ thresholds.

Table 5 shows some of the chronic-frequent patterns discovered in Shop-14 dataset at $minSup = 1\%$, $maxPrd = 5\%$ and $minPR = 1\%$. It can be observed that none of these patterns were appearing periodically throughout the database, however, there were periodically appearing in distinct subsets of the database. Using the approach discussed in [8], we have made an effort to find periodic-frequent patterns with $minSup = 1\%$ and $maxPrd = 5\%$. Unfortunately, no pattern was discovered at these threshold values. It because all frequent patterns have failed to reappear at very short intervals throughout the database. Thus, the proposed model was able to discover useful patterns.

Table 5. The chronic-frequent patterns discovered in Shop-14 dataset

Chronic-frequent patterns	Range of *tids* containing the pattern
{{TV's}, {Analog camcorders}}	[9,4447], [6591,15843], [16964,25508][26649,32654]
{{Speakers for home cinemas}, {Home cinema systems-components}}	[18, 5970], [7971, 11473], [18905, 24096]
{{Washer dryers}, {Refrigerators, freezers, show cases}, {built-in ovens, hobs, grills}}	[4,4655], [13824, 19589], [40232, 45721]
{{Built-in dish washers}, {Refrigerators, freezers, show cases}}	[13639,19544], [48495, 53310]

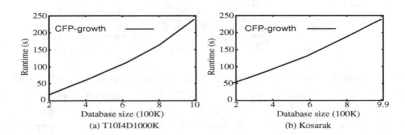

Fig. 5. Scalability of CFP-growth. (a) T10I4D1000K dataset and (b) Kosarak dataset.

5.2   The Scalability Test

We study the scalability of our CFP-growth algorithm on execution time by varying the number of transactions in $T10I4D1000K$ and $Kosarak$ datasets. In the literature, these two datasets were widely used to study the scalability of algorithms [23, 8]. The experimental setup was as follows. Each dataset was divided into five portions with 0.2 million transactions in each part. Then we investigated the performance of CFP-growth after accumulating each portion with previous parts. For each experiment, we set $minSup = 10\%$, $maxPrd = 1\%$ and $minPR = 10\%$.

Figure 5 (a) and (b) respectively show the runtime requirements of CFP-growth on the T10I4D1000K and Kosark datasets with the increase of dataset size. It is clear from the graphs that as the database size increases, overall tree construction and mining time increases. However, CFP-growth shows stable performance of about linear increase of runtime with respect to the database size. Therefore, it can be observed from the scalability test that CFP-growth can mine the patterns over large databases and distinct items with considerable amount of runtime.

6 Conclusions and Future Work

We introduced a new class of user-interest-based patterns known as chronic-frequent patterns. We also proposed a model for discovering such patterns. A highly compact tree structure, called CFP-tree, was proposed to capture the database contents in a compact form. A pattern-growth technique to discover the complete set of chronic-frequent patterns from CFP-tree has been introduced. The experimental results suggest that our CFP-growth can be runtime efficient, and highly scalable as well.

As a part of future work, we would like to extend our work to improve the performance of association rule-based recommender systems. Furthermore, it is interesting to investigate the chronic behavior of the patterns in time-series databases, sequential databases, and data streams.

References

1. Han, J., Gong, W., Yin, Y.: Mining segment-wise periodic patterns in time-related databases. In: KDD, pp. 214–218 (1998)
2. Han, J., Dong, G., Yin, Y.: Efficient mining of partial periodic patterns in time series database. In: ICDE, pp. 106–115 (1999)
3. Yang, R., Wang, W., Yu, P.S.: Infominer+: mining partial periodic patterns with gap penalties. In: ICDM, pp. 725–728 (2002)
4. Berberidis, C., Vlahavas, I., Aref, W., Atallah, M., Elmagarmid, A.: On the discovery of weak periodicities in large time series. In: PKDD, pp. 51–61 (2002)
5. Cao, H., Cheung, D.W., Mamoulis, N.: Discovering partial periodic patterns in discrete data sequences. In: Dai, H., Srikant, R., Zhang, C. (eds.) PAKDD 2004. LNCS (LNAI), vol. 3056, pp. 653–658. Springer, Heidelberg (2004)
6. Aref, W.G., Elfeky, M.G., Elmagarmid, A.K.: Incremental, online, and merge mining of partial periodic patterns in time-series databases. IEEE TKDE 16(3), 332–342 (2004)
7. Agrawal, R., Imieliński, T., Swami, A.: Mining association rules between sets of items in large databases. In: SIGMOD, pp. 207–216 (1993)
8. Tanbeer, S.K., Ahmed, C.F., Jeong, B.-S., Lee, Y.-K.: Discovering periodic-frequent patterns in transactional databases. In: Theeramunkong, T., Kijsirikul, B., Cercone, N., Ho, T.-B. (eds.) PAKDD 2009. LNCS, vol. 5476, pp. 242–253. Springer, Heidelberg (2009)
9. Kiran, R.U., Reddy, P.K.: Towards efficient mining of periodic-frequent patterns in transactional databases. DEXA (2) 2, 194–208 (2010)

10. Surana, A., Kiran, R.U., Reddy, P.K.: An efficient approach to mine periodic-frequent patterns in transactional databases. In: Cao, L., Huang, J.Z., Bailey, J., Koh, Y.S., Luo, J. (eds.) PAKDD Workshops 2011. LNCS, vol. 7104, pp. 254–266. Springer, Heidelberg (2012)

11. Amphawan, K., Lenca, P., Surarerks, A.: Mining top-K periodic-frequent pattern from transactional databases without support threshold. In: Papasratorn, B., Chutimaskul, W., Porkaew, K., Vanijja, V. (eds.) IAIT 2009. CCIS, vol. 55, pp. 18–29. Springer, Heidelberg (2009)

12. Han, J., Cheng, H., Xin, D., Yan, X.: Frequent pattern mining: Current status and future directions. DMKD 14(1) (2007)

13. Antunes, C.M., Oliveira, A.L.: Temporal data mining: An overview. In: Workshop on Temporal Data Mining, KDD (2001)

14. Özden, B., Ramaswamy, S., Silberschatz, A.: Cyclic association rules. In: ICDE, pp. 412–421 (1998)

15. Chen, S.-S., Huang, T.C.-K., Lin, Z.-M.: New and efficient knowledge discovery of partial periodic patterns with multiple minimum supports. J. Syst. Softw. 84(10), 1638–1651 (2011)

16. Kiran, R.U., Reddy, P.K.: An alternative interestingness measure for mining periodic-frequent patterns. In: Yu, J.X., Kim, M.H., Unland, R. (eds.) DASFAA 2011, Part I. LNCS, vol. 6587, pp. 183–192. Springer, Heidelberg (2011)

17. Kiran, R.U., Kitsuregawa, M.: Discovering quasi-periodic-frequent patterns in transactional databases. In: Bhatnagar, V., Srinivasa, S. (eds.) BDA 2013. LNCS, vol. 8302, pp. 97–115. Springer, Heidelberg (2013)

18. Han, J., Pei, J., Yin, Y., Mao, R.: Mining frequent patterns without candidate generation: A frequent-pattern tree approach. Data Min. Knowl. Discov. 8(1), 53–87 (2004)

19. Zaki, M.J., Hsiao, C.-J.: Efficient algorithms for mining closed itemsets and their lattice structure. IEEE Trans. on Knowl. and Data Eng. 17(4), 462–478 (2005)

20. Zhi-jun, X., Hong, C., Li, C.-P.: An efficient algorithm for frequent itemset mining on data streams. In: Perner, P. (ed.) ICDM 2006. LNCS (LNAI), vol. 4065, pp. 474–491. Springer, Heidelberg (2006)

21. Weblog dataset, http://web.archive.org/web/20070713202946rn_1/lisp.vse.cz/challenge/CURRENT/

22. Zheng, Z., Kohavi, R., Mason, L.: Real world performance of association rule algorithms. In: KDD 2001, pp. 401–406 (2001)

23. Kiran, R.U., Reddy, P.K.: An alternative interestingness measure for mining periodic-frequent patterns. DASFAA 1, 183–192 (2011)

High Utility Rare Itemset Mining
over Transaction Databases

Vikram Goyal[1], Siddharth Dawar[1], and Ashish Sureka[2]

[1] Indraprastha Institute of Information Technology-Delhi (IIIT-D), India
{vikram,siddharthd}@iiitd.ac.in
[2] Software Analytics Research Lab (SARL), India
ashish@iiitd.ac.in

Abstract. High-Utility Rare Itemset (HURI) mining finds itemsets from
a database which have their utility no less than a given minimum utility
threshold and have their support less than a given frequency threshold.
Identifying high-utility rare itemsets from a database can help in better
business decision making by highlighting the rare itemsets which give
high profits so that they can be marketed more to earn good profit.
Some two-phase algorithms have been proposed to mine high-utility rare
itemsets. The rare itemsets are generated in the first phase and the high-
utility rare itemsets are extracted from rare itemsets in the second phase.
However, a two-phase solution is inefficient as the number of rare itemsets is enormous as they increase at a very fast rate with the increase in
the frequency threshold. In this paper, we propose an algorithm, namely
UP-Rare Growth, which uses UP-Tree data structure to find high-utility
rare itemsets from a transaction database. Instead of finding the rare
itemsets explicitly, our proposed algorithm works on both frequency and
utility of itemsets together. We also propose a couple of effective strategies to avoid searching the non-useful branches of the tree. Extensive
experiments show that our proposed algorithm outperforms the state-of-
the-art algorithms in terms of number of candidates.

Keywords: Data Mining, Pattern Mining, Rare Itemset Mining, Rare
Utility Itemset, Utility Mining.

1   Introduction

High-Utility Rare Itemset mining finds those itemsets from the database which
are rare as well as of high utility. An itemset is defined as a high utility itemset
if its utility value is no less than a given minimum utility threshold. The utility
of an itemset is a function of its quantity and the profit value associated with
it. An itemset is rare if its support is no greater than a given maximum support
threshold. Mining high-utility rare itemsets (HURI) [1] from a database may be
interesting for business organizations. For example, identifying HURI in a retail
store will help the retail owner to focus on items that should be marketed well to
earn more profit. High utility rare itemset mining also finds its use in applications
of anomaly detection such as identifying fraudulent credit card transactions,

W. Chu et al. (Eds.): DNIS 2015, LNCS 8999, pp. 27–40, 2015.
© Springer International Publishing Switzerland 2015

medicine [2], molecular biology [3] and security [4]. A lot of work has been done in the areas of Frequent Itemset Mining (FIM) [5–7] and high-utility itemset mining [8–11] separately. FIM mines those itemsets from the database which are frequent without considering the profit value or quantity value associated with the items. High utility itemset mining removes this limitation, but still does not consider the frequency of itemsets into account. Itemsets which are interesting in utility as well as frequency aspects may not be identified if only utility or frequency objective is considered.

In this paper, we focus on mining high-utility rare itemsets from transaction databases. Jyothi et al. [1] proposed a two-phase algorithm to find high utility rare itemsets from transaction databases. The rare itemsets are mined in the first phase and utility of rare itemsets are computed in the next phase to find high utility rare itemsets. Jyothi et al. [12] proposed an approach similar to their previous work [1] for finding profitable transactions along with high utility rare itemsets from a transaction database. However, the two-phase approach to mine high utility rare itemsets is not efficient as the amount of rare itemsets increase rapidly with the increase in frequency threshold resulting in longer execcution time.

We propose an algorithm called, UP-Rare Growth, which uses a UP-Tree data structure [13] to find high utility rare itemsets. Our proposed algorithm works on both utility and frequency dimensions together and generates candidate high utility rare itemsets in the first phase which are then verified in the second phase. Our approach for high-utility rare itemset mining is efficient because of following reasons:

1. UP-Tree allows for a compressed representation of the database and allow to develop an efficient algorithm which takes both frequency and utility dimensions simultaneously into account,
2. Our approach is a pattern-growth approach which allows for the generation of a significantly lesser number of candidates as compared to a level-wise approaches like Apriori [5].

Our novel research contributions can be summarized as follows:

1. We propose an efficient algorithm, UP-Rare growth, to find high-utility-rare itemsets from a transaction database.
2. We propose effective pruning strategies which help in computation of results faster by pruning the non-promising search space.
3. We conduct extensive experiments on Mushroom dataset to show that our proposed algorithm outperforms state-of-the-art algorithms in terms of the number of candidates.

2   Related Work

Frequent-itemset mining [5–7] has been studied extensively in the literature. Agrawal et al. [5] proposed an algorithm named Apriori, for mining association

rules from market-basket data. Their algorithm was based on the downward closure property [5]. The downward closure property states that every subset of a
frequent itemset is also frequent. Park et al. [14] proposed a hash based algorithm
for mining association rules which generates less number of candidates compared
to Apriori algorithm. Zaki et al. [15] proposed an algorithm, namely ECLAT,
for mining association rules which used itemset clustering to find the set of potentially maximal frequent itemsets. Han et al. [6] proposed a pattern-growth
algorithm to find frequent itemsets by using FP-tree data structure. Other variants of itemset mining problem that have been proposed in the literature are
high utility itemset mining, high utility-frequent itemset mining and high-utility
rare itemset mining. However, frequent-itemset mining algorithms can't be used
to find high utility itemsets as it is not necessarily true that a frequent itemset
is also a high utility itemset in the database. On the other hand, mining high-
utility patterns is challenging compared to the frequent-itemset mining, as there
is no downward closure property [5], like we have in frequent-itemset mining
scenario.

Several algorithms have also been proposed to find high utility itemsets. Liu
et al.[10] proposed a two-phase algorithm which generates candidate high utility
itemsets in the first phase and verification is done in the second phase. Ahmed
et al.[16] proposed another two-phase algorithm, which uses a data structure
named IHUP-Tree, to mine high utility patterns incrementally from dynamic
databases. The problem with the above mentioned algorithms is the generation
of a huge amount of candidates in the first phase which leads to longer execution
times. In order to reduce the number of candidates, Tseng et al.[13] proposed
a new data structure called UP-Tree and algorithms, namely UP-Growth [13]
and UP-Growth+ [9]. The authors proposed effective strategies like DGU, DGN,
DLU and DLN to compute better utility estimates.

Some work has also been done on high frequency-high utility [17] and high
utility-rare itemset [1], [12]. Yeh et al. [17] proposed a bottom-up and top-down
two phase algorithms to find frequent high utility itemsets. They introduced
the concept of quasi-utility-frequency which is upward closed with respect to
the lattice of all itemsets. The top-down algorithm finds quasi-utility-frequency
candidates in the first phase, which are verified in the second phase. The problem
of finding rare itemsets have been investigated by some authors [18], [19], [20].
Koh et al. [18] proposed an algorithm Apriori-Inverse for discovering sporadic
rules by discarding all the itemsets which have their support greater than the
maximum frequency threshold. Troiano et al. [20] proposed a top-down algorithm
which used power set lattice to find rare itemsets. Pillai et al. [1] proposed an
algorithm HURI for finding high utility rare itemsets. Their proposed algorithm
used the concept of Aprori-Inverse. Pillai et al. [12] proposed a modified HURI
algorithm to find profitable transactions which contained rare itemsets and the
share of such items in the overall profit of transactions. However, the above
mentioned algorithm generates rare itemsets in the first phase, which are verified
in the second phase.

Table 1. *Example Database*

TID	Transaction	TU
T_1	$(C:5)\,(D:20)$	70
T_2	$(C:1)\,(F:40)$	42
T_3	$(A:1)\,(B:1)\,(C:2)\,(G:10)$	20
T_4	$(A:1)\,(B:1)\,(C:2)$	10
T_5	$(A:5)\,(C:10)$	45
T_6	$(B:1)\,(C:1)\,(E:1)$	5
T_7	$(B:1)\,(C:1)\,(E:1)\,(G:10)$	15
T_8	$(B:1)\,(C:1)\,(E:1)\,(H:1)$	6
T_9	$(C:10)\,(E:10)$	40
T_{10}	$(A:1)\,(B:1)\,(C:1)$	8

Table 2. *Profit Table*

Item	A	B	C	D	E	F	G	H
Profit	5	1	2	3	2	1	1	1

3   Background

In this section, we present some definitions given in the earlier works and describe the problem statement formally. We also discuss the UP-Tree data structure briefly.

3.1   Preliminary

We have a set of m distinct items $I = \{i_1, i_2, ..., i_m\}$, where each item has a profit $pr(i_p)$ (*external utility*) associated with it. An itemset X of length k is a set of k items $X = \{i_1, i_2, ..., i_k\}$, where for $j \in 1.....k$, $i_j \in I$. A transaction database $D = \{T_1, T_2, ..., T_n\}$ consists of a set of n transactions, where every transaction has a subset of items belonging to I. Every item I_p in a transaction T_d has a quantity $q(i_p, T_d)$ associated with it.

Definition 1. *The utility of an item I_p in a transaction T_d is the product of the profit of the item and its quantity in the transaction i.e. $u(i_p, T_d) = q(i_p, T_d) * pr(i_p)$.*

Definition 2. *The utility of an itemset X in a transaction T_d is denoted as $u(X, T_d)$ and defined as $\sum_{X \subseteq T_d \wedge i_p \in X} u(i_p, T_d)$.*

Definition 3. *The utility of a transaction T_d is denoted as $TU(T_d)$ and defined as $\sum_{i_p \in T_d} u(i_p, T_d)$.*

Let us consider the example database shown in Table 1 and the profit values in Table 2. The utility of item $\{A\}$ in $T_3 = 1 \times 5 = 5$ and the utility of itemset $\{A, B\}$ in T_3 denoted by $u(\{A, B\}, T_3) = u(A, T_3) + u(B, T_3) = 5 + 1 = 6$.

Definition 4. *The utility of an itemset X in database D is denoted as $u(X)$ and defined as $\sum_{X \subseteq T_d \wedge T_d \in D} u(X, T_d)$.*

For example, $u(A,B) = u(\{A,B\},T_3) + u(\{A,B\},T_4) + u(\{A,B\},T_{10}) = 7 + 7 + 7 = 21$.

Definition 5. *An itemset is called a high utility itemset if its utility is no less than a user-specified minimum threshold denoted by min_util.*

For example, $u(A,C) = u(\{A,C\},T_3) + u(\{A,C\},T_4) + u(\{A,C\},T_5) + u(\{A,C\},T_{10}) = 9 + 9 + 45 + 7 = 70$. If $min\_util = 30$, then $\{A,C\}$ is a high utility itemset. However, if $min\_util = 75$, then $\{A,C\}$ is a low utility itemset.

Table 3. *Rare Itemset Table*

Itemsets	List of rare itemsets
1-itemset	{ D }, { F }, { H }
2-itemset	{ AG }, { BH }, { CD }, { CF }, { CH }, { EG }, { EH }
3-itemset	{ ABG }, { ACG }, { BCH },{ BEH }, { BEG }, { CEG }, { CEH }
4-itemset	{ ABCG }, { BCEG }, { BCEH }

Definition 6. *The support of an itemset X denoted by $sup(X)$ is the number of transactions in database D which contain itemset X.*

For example, the $sup(\{A,C\}) = 4$.

Definition 7. *An itemset X is called a rare itemset, if $sup(X) < max\_sup\_threshold$.*

Let $max\_sup\_threshold = 2$. The rare itemsets are shown in Table 3.

Table 4. *Rare High Utility Itemset Table*

Itemsets	List of high utility rare itemsets
1-itemset	{D}, { F }
2-itemset	{ CD }, { CF }
3-itemset	{∅}
4-itemset	{∅}

Problem Statement. Given a transaction database D, a minimum utility threshold min_util and maximum support threshold $max\_sup\_threshold$, the aim is to find all the itemsets which are rare as well as of high utility, i.e. itemsets which have utility no less than min_util and support value less than $max\_sup\_threshold$.

Table 5. *MIU Table*

Item	A	B	C	D	E	F	G	H
MIU	5	2	2	60	2	40	10	1

Let $min\_util = 30$ and $max\_sup\_threshold = 2$. The set of high utility rare itemsets are shown in Table 4.

We will now describe the concept of transaction utility and transaction weighted downward closure(TWDC)[8].

Definition 8. *The transaction utility of a transaction T_d is denoted by $TU(T_d)$ and defined as $u(T_d, T_d)$.*

For example, the transaction utility of every transaction is shown in Table 1.

Definition 9. *Transaction-weighted utility of an itemset X is the sum of the transaction utilities of all the transactions containing X, which is denoted as $TWU(X)$ and defined as $\sum_{X \subseteq T_d \wedge T_d \in D} TU(T_d)$.*

Definition 10. *An itemset X is called a high-transaction-weighted utility itemset (HTWUI), if $TWU(X)$ is no less than min_util.*

Property 1 (Transaction-weighted downward closure). For any itemset X, if X is not a (HTWUI), any superset of X is not a HTWUI.
For example, $TU(T_1) = u(\{CD\}, T_1) = 70; TWU(\{A\}) = TU(T_3) + TU(T_4) + TU(T_5) + TU(T_{10}) = 83$. If $min\_util = 80$, $\{A\}$ is a HTWUI. However, if $min\_util = 100$, $\{A\}$ and any of its supersets are not HTWUIs.

We will now describe the concepts [9] for computing ovestimated utility of an item.

Definition 11. *Minimum item utility of item i_p in database D, denoted as $miu(i_p)$ is $i_p's$ utility in transaction T_d if there does not exist a transaction T'_d such that $u(i_p, T'_d) < u(i_p, T_d)$.*

The minimum item utility of the items in database D is shown in the Table 5

Definition 12. *Assume that N_x is the node which records the item x in the path p in a UP-Tree and N_x is composed of items x from the set of transactions $TIDSET(T_X)$. The minimum node utility of x in p is denoted as $mnu(x, p)$ and defined as $min_{\forall T \in TIDSET(T_X)}(u(x, T))$.*

3.2   UP-Tree

Each node N in UP-Tree [13] consists of a name $N.item$, overestimated utility $N.nu$, support count $N.count$, a pointer to the parent node $N.parent$ and a pointer $N.hlink$ to the node which has the same name as $N.name$. The root of

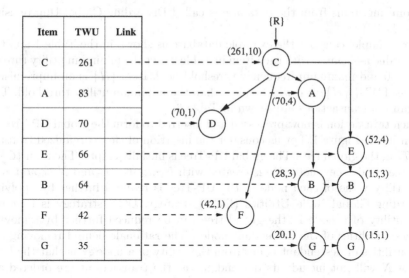

Item	TWU	Link
C	261	
A	83	
D	70	
E	66	
B	64	
F	42	
G	35	

Fig. 1. Global UP-Tree

the tree is a special empty node which points to its child nodes. The support count of a node N along a path is the number of transactions contained in that path that have the item $N.item$. $N.nu$ is the overestimated utility of an itemset along the path from node N to the root. In order to facilitate efficient traversal, a header table is also maintained. The header table has three columns, *Item*, *TWU* and *Link*. The nodes in a UP-Tree along a path are maintained in descending order of their TWU values. All nodes with the same label are stored in a linked list and the link pointer in the header table points to the head of the list.

4   Mining High Utility Rare Itemsets

In this section, we will describe our algorithm UP-Rare Growth for mining high utility rare itemsets. We will illustrate the working of our algorithm with an example and will formally prove its correctness.

4.1   Construction of a Global UP-Tree

In this subsection, we will discuss how to construct the global UP-Tree from the database D. The global UP-Tree is constructed in two scans of the database. In the first scan, the TWU value of every item is computed. The unpromising items are removed from the transaction database and transactions are reorganized in decreasing order of their TWU values. Unpromising items are the items which have their TWU value less than the minimum utility threshold. The removal of unpromising items from the database is called Discarding Global Unpromising items (DGU).

For example, consider the example database as shown in the Table 1 and the profit value associated with each item in Table 2. Let the minimum utility threshold be 30 and maximum frequency threshold be 2. Item $\{H\}$ is an unpromising item as TWU($\{H\}$) is 6 which is less than the minimum utility threshold. The reorganized transactions are shown in Table 6.

Each transaction is now processed and inserted to form the global UP-Tree as shown in the Figure 1. Let us consider the insertion of the reorganized transaction T_1' in the global tree. The global UP-Tree is initially empty. The item $\{C\}$ is processed and a new node N_C is created with $N_C.item = C$ and $N_C.count = 1$. The utility value of each node in the UP-Tree is reduced further by applying Discarding Global Node Utilities (DGN) strategy. DGN strategy is that the node utility of a node in the global tree can be reduced further by removing the node utilities of the descendant nodes. The rationale behind removing the node utilities of descendant nodes from the utility of a node N is that the local tree of N will not include its descendants as the transactions are ordered according to TWU values. The utility of node N_C is computed by subtracting the utilities of its descendants, i.e. the items after C in the reorganized transaction. In our example, $N_C.nu=RTU(T_1') - u(\{D\}, T_1')$=70-60=10. Next, item $\{D\}$ is processed. A new node N_D is created with $N_D.item = D$ and $N_D.count = 1$. Since, there is no item after $\{D\}$ in the transaction T_1', its utility is equal to $RTU(T_1')$ i.e. 70. If the node of the item to be inserted in the tree is already present along that path, the support count and node utilities are simply incremented. Similarly, other reorganized transactions are inserted to construct the global UP-Tree. The strategies DLU and DLN are similar to DGU and DGN, but are applied to the local UP-Tree. Since, exact utilities are not stored in the global UP-Tree, utilities of unpromising items are estimated using the minimum item utility and minimum node utility as per Definition 11 and 12.

Table 6. *Reorganized Transactions*

TID	Reorganized Transaction	RTU
T_1'	$(C:5)\,(D:20)$	70
T_2'	$(C:1)\,(F:40)$	42
T_3'	$(C:2)\,(A:1)\,(B:1)\,(G:10)$	20
T_4'	$(C:2)\,(A:1)\,(B:1)$	10
T_5'	$(C:10)\,(A:5)$	45
T_6'	$(C:1)\,(E:1)\,(B:1)$	5
T_7'	$(C:1)\,(E:1)\,(B:1)\,(G:10)$	15
T_8'	$(C:1)\,(E:1)\,(B:1)$	5
T_9'	$(C:10)\,(E:10)$	40
T_{10}'	$(C:1)\,(A:1)\,(B:1)$	8

4.2   UP-Rare Growth

The algorithm UP-Rare Growth takes as input a UP-Tree, a header table, an itemset, a minimum utility threshold, a maximum support threshold and returns the candidate rare high utility itemsets. The steps of the algorithm are shown in Algorithm 1. The algorithm starts with an empty prefix and extends the prefix with item i_k of the header table. In this process of extension (growth), a conditional pattern base(CPB) is constructed from the prefix. The CPB of the prefix extended with item i_k consists of all the paths through which i_k is reachable from the root of the tree. A local UP-Tree is constructed and strategies DLU and DLN are applied. We apply the following strategies while processing for a prefix X just extended with item i_k to prune the search space:

1. If X has low TWU i.e. estimated utility of any itemset containing X is less than the minimum utility threshold, X prefix is not processed further and the algorithm proceeds with the next alternative of the header table. Else, X is processed further.
2. If the support count of every leaf node of the UP-Tree is greater than the given support threshold, it is guaranteed that no rare itemset can be found using item i_k as a prefix. In this case, prefix i_k is not processed further.

Algorithm 1. UP-Rare Growth(T_x, H_x, X)

Input: A UP-Hist tree T_x, a header table H_x for T_x, an itemset X, a minimum utility threshold min_util and maximum support threshold $max\_sup_t hreshold$.
Output: All candidate rare High Utility Itemsets in T_x.

1: **for** entry i_k in H_x **do**
2:     Traverse the linked list associated with i_k and accumlate sum of node utilities $nu\_sum(i\_k)$.
3:     **if** $nu\_sum(X) \geq min\_util$ **then**
4:         **if** (**then** $\sup(i\_k) < max\_sup\_threshold$))
5:             Consider $Y = X \cup i\_k$ as a candidate and construct CPB of Y.
6:         **else**
7:             Construct the CPB of Y.
8:         **end if**
9:         Put local promising items in $Y - CPB$ into H_Y and apply DLU to reduce path utilities.
10:        Insert every reorganized path into T_Y after applying DLN.
11:        **if** $T_Y \neq null$ **then**
12:            **if** support of every leaf node of $T\_Y > max\_sup\_threshold$ **then**
13:                continue
14:            **else**
15:                Call UP-Rare Growth(T_Y, H_Y, Y)
16:            **end if**
17:        **end if**
18:    **end if**
19: **end for**

Now, we will illustrate the working of our algorithm with an example. Consider the global UP-Tree shown in the Figure 1. Let the minimum utility threshold be 30 and maximum frequency support be 2. The algorithm picks the lowest entry from the header table i.e. G and accumulates its node utility. Since the accumulated node utility $nu_{sum}(G)$ i.e. 35 is greater than the maximum utility threshold, item $\{G\}$ will be processed further. However, $\{G\}$ is not added to the set of candidate itemsets as the sup(G) is not less than the maximum frequency threshold. The conditional pattern base of $\{G\}$ is constructed as shown in the Table 7. The TWU values of the items in the CPB of $\{G\}$ are computed and the

Table 7. $\{G\} - CPB$ after applying DGU DGN and DLN

Retrieved Path: Path utility	Reorganized Path: Path utility (after DLU)	Support
< CAB >: 20	< CB >: 15	1
< CEB >: 15	< CB >: 13	1

unpromising items are removed to get the reorganized paths. In the CPB of $\{G\}$, Item $\{A\}$ and $\{E\}$ are unpromising as $TWU(\{A\}) = 20$ and $TWU(\{E\}) = 15$ is less than the minimum utility threshold. The reorganized path utilities are computed using minimum node utilities similar to UP-Growth+ i.e.,
$pu(< CAB >, \{G\} - CPB) = 20$ - $A.mnu \times < CAB >.support=20-5\times1=15$.
$pu(< CEB >, \{G\} - CPB) = 15$ - $E.mnu \times < CEB >.support=15-2\times1=13$.
The algorithm is called recusively for the itemset $\{GC\}$, $\{GB\}$ and $\{GCB\}$. However, all the itemsets which have G as a prefix are low utility itemsets as their TWU value is less than the minimum utility threshold. Similarly, the next item $\{F\}$ is processed from the header table.

Table 8. $\{B\} - CPB$ after applying DGU DGN and DLN

Retrieved Path: Path utility	Reorganized Path: Path utility (after DLU)	Support
< CA >: 28	< CA >: 28	3
< CE >: 15	< CE >: 15	3

We will now focus on the processing of item $\{B\}$ in the global header table. The linked list associated with B is traversed from the header table and the sum of node utilities is accumulated. Since the TWU of $\{B\}$ is greater than the minimum utility threshold, it is processed further. However, $\{B\}$ is not added to the candidates as sup(B) is 6 which is greater than the maximum frequency threshold. The conditional pattern base of $\{B\}$ is constructed as shown in the Table 8. There are no unpromising items in the CPB of $\{B\}$ and a local UP-Tree is constructed as shown in the Figure 2. However, there is no rare itemset containing $\{B\}$ as the support of every leaf node in the local UP-Tree is greater than the maximum frequency threshold. After processing the remaining items in the global header table, the complete set of candidate high utility rare itemsets are generated. The candidate itemsets obtained from UP-Rare Growth are $\{F\}$, $\{CF\}$, $\{D\}$ and $\{CD\}$. The generated candidates are verified by scanning the original database again and computing the exact utilities of the candidate itemsets.

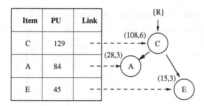

Fig. 2. Local UP-Tree

Claim 1. *Algorithm UP-Rare Growth does not generate any false negatives.*

Proof. The algorithm UP-Rare Growth prunes an itemset and its supersets on the basis of two rules: (1) If the overestimated utility of an itemset is less than the minimum utility threshold, (2) the support count of all leaf nodes in the tree constructed from the CPB of the itemset, is greater than the maximum support threshold. We know that the TWU value associated with every itemset is an upper bound on the exact utility value and satisfies the downward closure property. Therefore, if an itemset is marked off as low utility, it is guaranteed that this itemset and its supersets will be of low utility. So, rule 1 will not generate any false negatives. The algorithm also prunes the supersets of an itemset if all the leaf nodes have their support count greater than the maximum frequency threshold. Since the transactions are reorganized in the decreasing order of TWU values, the leaf nodes have the least support value compared to its ancestors in the tree. Therefore, if the support of all the leaf nodes of the UP-Tree constructed from the conditional pattern base of an itemset is greater than the maximum support threshold, it is guaranteed that there can't be any superset of that itemset which is rare. This proves the claim. □

5 Experiments and Results

In this section, we compare the performance of our proposed algorithm UP-Rare Growth against the state-of-the-art algorithm HURI [1]. We implemented all the algorithms in Java on Eclipse 3.5.2 platform with JDK 1.6.0_24. The experiments were performed on an Intel Xeon(R) CPU=26500@2.00 GHz with 64 GB RAM. We ran our experiments on the Mushroom dataset which was obtained from FIMI repository [21]. The quantity and external utility for the Mushroom datasets was generated using log-normal distribution.

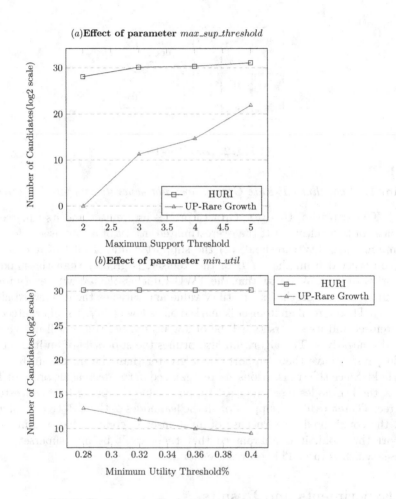

Fig. 3. Performance Evaluation on Mushroom dataset

We studied the impact of the parameters, maximum support threshold and minimum utility threshold on the performance of the algorithms. We observed that there were no high-utility itemsets in our Mushroom database for the utility threshold 2,50,00,000. Therefore, we set this value as the maximum utility threshold and represent the different values of min_util as percent with respect to this value. We compare the performance of the algorithms in terms of the number of candidates generated after the first phase. The number of candidates is represented on a log scale with base 2. In order to study the effect of the maximum support threshold, we fixed $min\_util = 0.32\%$ and the results are shown in Figure 3(a).

We expect the number of rare itemsets to increase exponentially with varying support threshold and the results meet our expectation. The results show that our algorithm generates a significantly lesser number of candidates compared to HURI and the verification time taken by HURI will be very large as the verification time depends upon the number of candidates. In order to study the effect of minimum utility threshold, we fixed the maximum support threshold to 3 and the results are shown in Figure 3(b). The number of candidates generated by HURI remains constant with the varying minimum utility threshold as the algorithm doesn't take the utility dimension into account while computing the number of candidates. We also observe that the number of candidates generated by our algorithm decrease with an increase in the minimum utility threshold. The results clearly demonstrate the importance of taking the utility dimension into account when computing the candidate high-utility rare itemsets.

6 Conclusion and Future Work

In this paper, we proposed a novel algorithm UP-Rare Growth, for mining high-utility rare itemsets. Our algorithm considers both utility and frequency dimensions simultaneously and uses effective strategies to reduce the search space. Experimental results show that our proposed algorithm outperforms the state-of-the-art algorithm in terms of number of candidates.

References

1. Pillai, J., Vyas, O.P., Muyeba, M.: Huri–a novel algorithm for mining high utility rare itemsets. In: Advances in Computing and Information Technology, pp. 531–540. Springer, Heidelberg (2013)
2. Medici, F., Hawa, M.I., Giorgini, A.N.G.E.L.A., Panelo, A.R.A.C.E.L.I., Solfelix, C.M., Leslie, R.D., Pozzilli, P.: Antibodies to gad65 and a tyrosine phosphatase-like molecule ia-2ic in filipino type 1 diabetic patients. Diabetes Care 22(9), 1458–1461 (1999)
3. Shi, W., Ngok, F.K., Zusman, D.R.: Cell density regulates cellular reversal frequency in myxococcus xanthus. Proceedings of the National Academy of Sciences 93(9), 4142–4146 (1996)
4. Saha, B., Lazarescu, M., Venkatesh, S.: Infrequent item mining in multiple data streams. In: Seventh IEEE International Conference on Data Mining Workshops, ICDM Workshops 2007, pp. 569–574 (2007)

5. Agrawal, R., Ramakrishnan, Srikant, o.: Fast algorithms for mining association rules. In: Proc. 20th int. conf. very large data bases, VLDB, vol. 1215, pp. 487–499 (1994)
6. Han, J., Pei, J., Yin, Y.: Mining frequent patterns without candidate generation. In: ACM SIGMOD, vol. 29, pp. 1–12. ACM (2000)
7. Leung, C.K.-S., Khan, Q.I., Li, Z., Hoque, T.: Cantree: a canonical-order tree for incremental frequent-pattern mining. Knowledge and Information Systems 11(3), 287–311 (2007)
8. Liu, Y., Liao, W.-k., Choudhary, A.: A fast high utility itemsets mining algorithm. In: International Workshop on Utility-Based Data Mining, pp. 90–99. ACM (2005)
9. Tseng, V.S., Shie, B.-E., Wu, C.-W., Yu, P.S.: Efficient algorithms for mining high utility itemsets from transactional databases. IEEE Transactions on Knowledge and Data Engineering 25(8), 1772–1786 (2013)
10. Liu, Y., Liao, W.-k., Choudhary, A.K.: A two-phase algorithm for fast discovery of high utility itemsets. In: Ho, T.-B., Cheung, D., Liu, H. (eds.) PAKDD 2005. LNCS (LNAI), vol. 3518, pp. 689–695. Springer, Heidelberg (2005)
11. Shie, B.-E., Tseng, V.S., Yu, P.S.: Online mining of temporal maximal utility itemsets from data streams. In: ACM Symposium on Applied Computing, pp. 1622–1626. ACM (2010)
12. Pillai, J., Vyas, O.P.: Transaction profitability using huri algorithm [tphuri]. International Journal of Business Information Systems 2(1) (2013)
13. Tseng, V.S., Wu, C.-W., Shie, B.-E., Yu, P.S.: Up-growth: an efficient algorithm for high utility itemset mining. In: ACM SIGKDD, pp. 253–262. ACM (2010)
14. Park, J.S., Chen, M.-S., Yu, P.S.: An effective hash-based algorithm for mining association rules, vol. 24. ACM (1995)
15. Zaki, M.J., Parthasarathy, S., Ogihara, M., Wei, Li, o.: New algorithms for fast discovery of association rules. In: KDD, vol. 97, pp. 283–286 (1997)
16. Ahmed, C.F., Tanbeer, S.K., Jeong, B.-S., Lee, Y.-K.: Efficient tree structures for high utility pattern mining in incremental databases. IEEE Transactions on Knowledge and Data Engineering 21(12), 1708–1721 (2009)
17. Yeh, J.-S., Li, Y.-C., Chang, C.-C.: Two-Phase Algorithms for a Novel Utility-Frequent Mining Model. In: Washio, T., Zhou, Z.-H., Huang, J.Z., Hu, X., Li, J., Xie, C., He, J., Zou, D., Li, K.-C., Freire, M.M. (eds.) PAKDD 2007. LNCS (LNAI), vol. 4819, pp. 433–444. Springer, Heidelberg (2007)
18. Koh, Y.S., Rountree, N.: Finding sporadic rules using apriori-inverse. In: Ho, T.-B., Cheung, D., Liu, H. (eds.) PAKDD 2005. LNCS (LNAI), vol. 3518, pp. 97–106. Springer, Heidelberg (2005)
19. Szathmary, L., Napoli, A., Valtchev, P.: Towards rare itemset mining. In: 19th IEEE International Conference on Tools with Artificial Intelligence, ICTAI 2007, vol. 1, pp. 305–312. IEEE (2007)
20. Troiano, L., Scibelli, G., Birtolo, C.: A fast algorithm for mining rare itemsets. ISDA 9, 1149–1155 (2009)
21. Goethals, B., Zaki, M.J.: The fimi repository (2012)

Skyband-Set for Answering Top-k Set Queries of Any Users

Md. Anisuzzaman Siddique, Asif Zaman, and Yasuhiko Morimoto

Hiroshima University,
1-7-1 Kagamiyama, Higashi-Hiroshima, 739-8521, Japan
{siddique,d140094}@hiroshima-u.ac.jp, morimoto@mis.hiroshima-u.ac.jp

Abstract. Skyline computation fails to response variant queries that need to analyze not just individual object of a dataset but also their combinations. Therefore set skyline has attracted considerable research attention in the past few years. In this paper, we propose a novel variant of set skyline query called the "skyband-set" query. We consider a problem to select representative distinctive objectsets in a numerical database. Let s be the number of objects in each set and n be the total number of objects in the database. The number of objectsets in the database amounts to $_nC_s$. We propose an efficient algorithm to compute skyband-set of the $_nC_s$ sets where the cardinality of s varies from 1 to n. We investigate properties of skyband-set query computation and develop pruning strategies to avoid unnecessary objectset enumerations as well as comparisons among them. We conduct a set of experiments to show the effectiveness and efficiency of the propose algorithm.

Keywords: Skyline queries, Objectset, Skyband-set, Dominance relationship.

1 Introduction

A skyline query retrieves a set of objects, each of which is not dominated by any other objects. Consider an example in the field of financial investment: an investor tends to buy the stocks that can minimize the commission costs and predicted risks. As a result, the target can be formalized as finding the skyline stocks with minimum costs and minimum risks. Figure 1(a) shows seven stocks records with their costs (a_1) and risks (a_2). In the list the best choice for a client comes from the skyline, i.e., one of $\{O_1, O_2, O_3\}$ in general (see Figure 1(b)). Since the notion of the skyline operator [2] was introduced by Borzsonyi in 2001, it has attracted considerable attention due to its broad applications including product or restaurant recommendations [9], review evaluations with user ratings [8], querying wireless sensor networks [20], and graph analysis [22]. A number of efficient algorithms for computing skyline objects have been reported in the literature [4,7,12,21].

One of known weakness of the skyline query is that it can not answer various queries that require us to analyze not just individual object of a dataset but also

W. Chu et al. (Eds.): DNIS 2015, LNCS 8999, pp. 41–55, 2015.

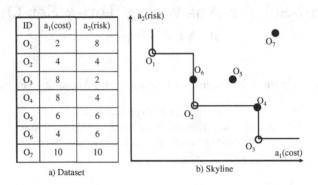

ID	a_1(cost)	a_2(risk)
O_1	2	8
O_2	4	4
O_3	8	2
O_4	8	4
O_5	6	6
O_6	4	6
O_7	10	10

a) Dataset

b) Skyline

Fig. 1. A Skyline Problem

their combinations or sets. It is very likely that an investor will not invest just in one stock, but in a combination of stocks which may allows the investor to obtain a lower investment and/or a lower risk. For example, investment in O_1 will render the lowest cost. However, this investment is also very risky. Are there any other stocks or sets of stocks which allow us to have a lower investment and/or a lower risk? These answers are often referred to as the investment portfolio. An investment portfolio can easily be found by applying objectset skyline query. Let s be the number of objects in each set and n be the number of objects in the dataset. The number of sets in the dataset amounts to $_nC_s$.

Assume an investor has to buy two stocks. Look at the example in Figure 1 again. The conventional skyline query outputs $\{O_1, O_2, O_3\}$ doesn't provide sufficient information for the objectset selection problem. Users may want to choose the portfolios which are not dominated by any others in order to minimize the total costs and the total risks. Figure 2(b) shows two objectsets consisting of stock records. Assume that their attribute values are the sums of their component values, and the objectset skyline problem is to find objectsets that have minimal values in attributes a_1 (cost) and a_2 (risk). Objectsets $\{O_{1,2}, O_{2,3}, O_{2,6}\}$ cannot be dominated by any other objectsets and thus they are the answer for the objectset skyline query when the objectset size s is equal to 2. There exists some efficient studies on the objectset skyline problem. The notion of the objectset skyline operator was introduced by Siddique et al. in 2010 [14]. They tried to find skyline objectsets that are on the convex hull enclosing all the objectsets. Su et al. proposed a solution to find the top-k optimal objectsets according to a user defined preference order of attributes [15]. Guo et al. proposed a pattern based pruning (PBP) algorithm to solve the objectsets skyline problem by indexing individuals objects [6].

The main obstacles of the objectsets skyline is that for any scoring function it is always retrieve top-1 objectset and sometimes retrieve too few objectsets as a query result. It can not solve the problem if an user wants more than one objectsets for a specific scoring function. To solve those problems in this paper, we propose a novel variant query of objectset skyline called "skyband-set" query.

Skyband-set query for K-skyband returns a set of objectsets, each objectset of which is not dominated by K other objectsets. In other words, an objectset is in the skyband-set query may be dominated by at most $K - 1$ other objectsets.

The skyband-set query helps us to retrieve desired objectsets without any scoring function. It also can increase the number of objectsets by increasing the skyband value of K. From skyband-set result an user can easily choose his/her desired objectsets by applying Top-k set queries. For the dataset in Figure 1, the skyband-set query for objectset size $s = 1$ and $K = 1$ retrieves objectsets $\{O_1, O_2, O_3\}$. For $s = 2$ and $K = 1$ it retrieves objectsets $\{O_{1,2}, O_{2,3}, O_{2,6}\}$ shown by double circles in Figure 2(b). Again, from Figure 1(b) for $s = 1$ and $K = 2$ we get objectsets $\{O_1, O_2, O_3, O_6\}$. For $s = 2$ and $K = 2$ skyband-set will retrieve $\{O_{1,2}, O_{1,3}, O_{1,6}, O_{2,3}, O_{2,5}, O_{2,6}, O_{3,4}\}$ (see Figure 2(b)). Therefore, one can use the skyband-set query at the preprocessing step for Top-k set query It is useful for candidate set generation to select his/her desired objectsets without any scoring function.

ID	a_1(cost)	a_2(risk)	ID	a_1(cost)	a_2(risk)
$O_{1,2}$	6	12	$O_{3,4}$	16	6
$O_{1,3}$	10	10	$O_{3,5}$	14	8
$O_{1,4}$	10	12	$O_{3,6}$	12	8
$O_{1,5}$	8	14	$O_{3,7}$	18	12
$O_{1,6}$	6	14	$O_{4,5}$	14	10
$O_{1,7}$	12	18	$O_{4,6}$	12	10
$O_{2,3}$	12	6	$O_{4,7}$	18	14
$O_{2,4}$	12	8	$O_{5,6}$	10	12
$O_{2,5}$	10	10	$O_{5,7}$	16	16
$O_{2,6}$	8	10	$O_{6,7}$	14	16
$O_{2,7}$	14	14			

a) Sets of 2 Stocks   b) Objectset skyband for $s = 2$

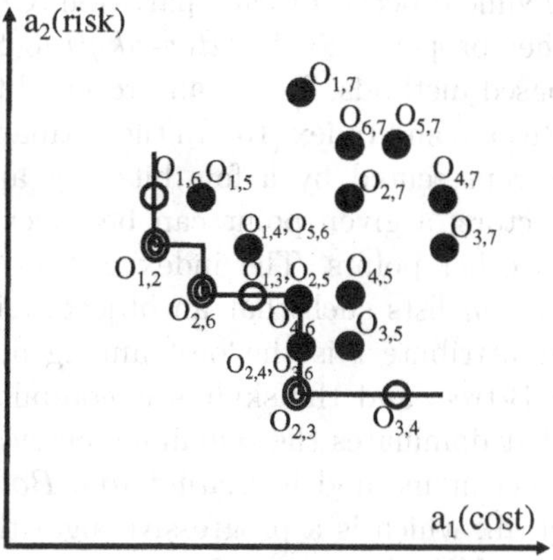

Fig. 2. An skyband-set Problem

The rest of this paper is organized as follows: Section 2 reviews related work. Section 3 presents the notions and properties for objectsets as well as the problem of skyband-set query. We provide detailed examples and analysis of propose algorithm for computing skyband-set in Section 4. We experimentally evaluate the proposed algorithm in Section 5 under a variety of settings. Finally, Section 6 concludes the paper.

2   Related Work

Our work is motivated by previous studies of skyline query processing as well as objectsets skyline query processing.

2.1   Skyline Query Processing

Borzsonyi et al. first introduced the skyline operator over large databases and proposed three algorithms: *Block-Nested-Loops(BNL)*, *Divide-and-Conquer* *(D&C)*, and B-tree-based schemes [2]. BNL compares each object of the database with every other object, and reports it as a result only if any other object does not dominate it. A window W is allocated in main memory, and the input relation is sequentially scanned. In this way, a block of skyline objects is produced in every iteration. In case the window saturates, a temporary file is used to store objects that cannot be placed in W. This file is used as the input to the next pass. *D&C* divides the dataset into several partitions such that each partition can fit into memory. Skyline objects for each individual partition are then computed by a main-memory skyline algorithm. The final skyline is obtained by merging the skyline objects for each partition. Chomicki et al. improved BNL by presorting, they proposed *Sort-Filter-Skyline(SFS)* as a variant of BNL [4]. Among index-based methods, Tan et al. proposed two progressive skyline computing methods Bitmap and Index [16]. In the Bitmap approach, every dimension value of a point is represented by a few bits. By applying bit-wise *AND* operation on these vectors, a given point can be checked if it is in the skyline without referring to other points. The index method organizes a set of m-dimensional objects into m lists such that an object O is assigned to list i if and only if its value at attribute i is the best among all attributes of O. Each list is indexed by a B-tree, and the skyline is computed by scanning the B-tree until an object that dominates the remaining entries in the B-trees is found. The current most efficient method is *Branch-and-Bound Skyline(BBS)*, proposed by Papadias et al., which is a progressive algorithm based on the *best-first nearest neighbor* *(BF-NN)* algorithm [12]. Instead of searching for nearest neighbor repeatedly, it directly prunes using the R*-tree structure.

Recently, more aspects of skyline computation have been explored. Chan et al. proposed k-dominant skyline and developed efficient ways to compute it in high-dimensional space [3]. Lin et al. proposed n-of-N skyline query to support online query on data streams, i.e., to find the skyline of the set composed of the most recent n elements. In the cases where the datasets are very large and stored distributedly, it is impossible to handle them in a centralized fashion [10]. Balke et al. first mined skyline in a distributed environment by partitioning the data vertically [1]. Vlachou et al. introduce the concept of extended skyline set, which contains all data elements that are necessary to answer a skyline query in any arbitrary subspace [18]. Tao et al. discuss skyline queries in arbitrary subspaces [17]. More skyline variants such as dynamic skyline [11] and reverse skyline [5] operators also have recently attracted considerable attention.

2.2   Objectsets Skyline Query Processing

Two related topics are "top-k combinatorial skyline queries" [15] and "convex skyline objectsets" [14]. Su et al. studied how to find top-k optimal combinations according to a given preference order in the attributes. Their solution is to retrieve non-dominated combinations incrementally with respect to the preference until the best k results have been found. This approach relies on the preference order of attributes and the limited number (top-k) of combinations queried. Both the preference order and the top-k limitation may largely reduce the exponential search space for combinations. However, in our problem there is no preference order nor the top-k limitation. Consequently, their approach cannot solve our problem easily and efficiently. Additionally, in practice it is difficult for the system or a user to decide a reasonable preference order. This fact will narrow down the applications of [15]. Siddique et al. studied the "convex skyline objectset" problem. It is known that the objects on the lower (upper) convex hull, denoted as CH, is a subset of the objects on the skyline, denoted as SKY. Every object in CH can minimize (maximize) a corresponding linear scoring function on attributes, while every object in SKY can minimize (maximize) a corresponding monotonic scoring function [2]. They aims at retrieving the objectsets in CH, however, we focuses on retrieving the objectsets in $CH \subseteq SKY$. Since their approach relies on the properties of the convex hull, it cannot extend easily to solve skyband-set query problem.

The similar related work is "Combination Skyline Queries" proposed in [6]. Guo et al. proposed a pattern based pruning (PBP) algorithm to solve the objectsets skyline problem by indexing individuals objects. The key problem of PBP algorithm is that it needs object selecting pattern in advance and the pruning capability depends on this pattern. If a wrong pattern choose at first this may largely increase the exponential search space for objectsets. Moreover, it fails to vary the cardinality of objectset size s. Our solution does not require to construct any pattern previously and also vary the objectset size s from 1 to n. There are some other works focusing on the combination selection problem but related to our work weakly [13,19]. Roy et al. studied how to select "maximal combinations". A combination is "maximal" if it exceeds the specified constraint by adding any new object. Finally, the k most representative maximal combinations, which contain objects with high diversities, are presented to the user. Wan et al. studies the problem to construct k profitable products from a set of new products that are not dominated by the products in the existing market [19]. They construct non-dominated products by assigning prices to the new products that are not given beforehand like the existing products.

However, none of the previous work do not focus on skyband-set query and are not suitable to solve this type query. Therefore in this studies we consider a new approach to solve the skyband-set query efficiently.

3 Preliminaries

Given a dataset D with m-attributes $\{a_1, a_2, \cdots, a_m\}$ and it has n objects $\{O_1, O_2, \cdots, O_n\}$, we use $O_i.a_j$ to denote the j-th dimension value of object O_i. Without loss of generality, we assume that smaller value in each attribute is better.

Definition *Dominance*: An object $O_i \in D$ is said to dominate another object $O_j \in D$, denoted as $O_i \leq O_j$, if $O_i.a_r \leq O_j.a_r$ $(1 \leq r \leq m)$ for all m attributes and $O_i.a_t < O_j.a_t$ $(1 \leq t \leq m)$ for at least one attribute. We call such O_i as *dominant object* and such O_j as *dominated object* between O_i and O_j.

Definition *Skyline*: An object $O_i \in D$ is said to be a *skyline object* of D, if and only if there does not exist any object $O_j \in D$ $(j \neq i)$ that dominates O_i, i.e., $O_j \leq O_i$ is not true. The skyline of D, denoted by $Sky(D)$, is the set of skyline objects in D. For dataset shown in Figure 1(a), object O_2 dominates $\{O_4, O_5, O_6, O_7\}$ and objects $\{O_1, O_3\}$ are not dominated by any other objects in D. Therefore, skyline query will retrieve $Sky(D) = \{O_1, O_2, O_3\}$ (see Figure 1(b)).

In the following, we first introduce the concept of objectset, and then use it to define skyband-set query. A s-objectset OS is made up of s objects selected from D, i.e., $OS = \{O_1, \cdots, O_s\}$ and for simplicity denoted as $OS = O_{1,\cdots,s}$. Each attribute value of OS is given by the formula below:

$$OS.a_j = f_j(O_1.a_j, \cdots, O_s.a_j), (1 \leq j \leq m) \tag{1}$$

where f_j is a monotonic aggregate function that takes s parameters and returns a single value. For the sake of simplicity, in this paper we consider that the monotonic scoring function returns the sum of these values; i.e.,

$$OS.a_j = \sum_{i=1}^{s} O_i.a_j, (1 \leq j \leq m) \tag{2}$$

though our algorithm can be applied on any monotonic aggregate function. Recall that the number of s-objectsets in D is $_nC_s = \frac{n!}{(n-s)!s!}$, we denote the number by $|S|$.

Definition *Dominance Relationship*: A s-objectset $OS \in D$ is said to dominate another s-objectset $OS' \in D$, denoted as $OS \leq OS'$, if $OS.a_r \leq OS'.a_r$ $(1 \leq r \leq m)$ for all m attributes and $OS.a_t < OS'.a_t$ $(1 \leq t \leq m)$ for at least one attribute. We call such OS as dominant s-objectset and OS' as dominated s-objectset between OS and OS'.

Definition *Objectset Skyline*: A s-objectset $OS \in D$ is said to be a skyline s-objectset if OS is not dominated by any other s-objectsets in D. The skyline of s-objectsets in D, denoted by $Sky_s(D)$, is the set of skyline s-objectsets in D. Assume $s = 2$, then for the dataset shown in Figure 2(a), 2-objectset $O_{1,2}, O_{2,3}$, and $O_{2,6}$ are not dominated by any other 2-objectsets in D. Thus, 2-objectset skyline query will retrieve $Sky_2(D) = \{O_{1,2}, O_{2,3}, O_{2,6}\}$ (see Figure 2(b)).

Definition *Skyband-set*: Skyband-set query returns a set of objectsets, each objectset of which is not dominated by K other objectsets. In other words, an objectset in the skyband-set query may be dominated by at most $K - 1$ other objectsets. If we want to apply skyband-set query in D and choose objectset size $s = 2$ and $K = 2$, then the skyband-set will retrieve $\{O_{1,2}, O_{1,3}, O_{1,6}, O_{2,3}, O_{2,5}, O_{2,6}, O_{3,4}\}$ as query result.

4 Objectsets Skyband Algorithm

In this section, we present our skyband-set method. It is an iterative algorithm also called level wise search. Initially, it computes 1-objectsets skyband; then all 2-objectsets skyband, and so on.

Initially, consider a skyband-set query where objectset size $s = 1$ and skyband size $K = 1$, that means we are seeking answer for normal skyline query and it will retrieve objects that are not dominated by other objects. Any conventional algorithm can be applied to answer this query. In this paper we use SFS method proposed in [4] to compute initial skyband-set query with $s = 1$ and $K = 1$. After performing domination check it will construct the following domination relation table called *domRelationTable* for objectset size $s = 1$.

Table 1. domRelationTable for 1-objectsets

Object	Dominant Object
O_1	$\varnothing$
O_2	$\varnothing$
O_3	$\varnothing$
O_4	O_2, O_3
O_5	O_2, O_6
O_6	O_2
O_7	$O_{1,\cdots,6}$

Table 1 shows that O_1, O_2, and O_3 are not dominated by other object thus for $s = 1$ and $K = 1$ skyband-set query result is $\{O_1, O_2, O_3\}$. If we keep objectset size $s = 1$ and increasing the skyband value of $K = 2$ then skyband-set result becomes $\{O_1, O_2, O_3, O_6\}$. Similarly, skyband-set query for $s = 1$ and $k = 3$ will retrieve $\{O_1, O_2, O_3, O_4, O_5, O_6\}$ and query for $s = 1$ and $K = 4$ will retrieve all objects as a result.

For the skyband-set query problem, the number of objectsets is $|S| = {}_nC_s$ for a dataset D containing n objects when we select objectsets of size s. This poses serious algorithmic challenges compared with the traditional skyline problem. As Figure 2(a) shows, $|S| = 21$ (${}_7C_2$) possible combinations are generated from seven objects when $s = 2$. Moreover to generate *domRelationTable* like Table 1 we need to perform domination check with 420 (21 * 20) comparisons.

Even for a small dataset with thousands of entries, the number of objectsets is prohibitively large and required huge number of comparisons. Thanks to the Theorem 1 and Theorem 2 which give us opportunity to receive objectsets dominance relationship without composing them. It also gives us opportunity to avoid many unnecessary comparisons required for domination check.

Theorem 1. *Suppose OS_1, OS_2, and OS_3 be the three objectsets in D. If objectset $OS_1 \leq OS_2$, then their super objectset with OS_3 also dominates, i.e., $OS_1OS_3 \leq OS_2OS_3$ is true.*

Proof: Suppose $OS_1OS_3 \leq OS_2OS_3$ is not true. After eradicate OS_3 from both objectsets we get $OS_1 \leq OS_2$, which contradict our assumption. Thus, if $OS_1 \leq OS_2$ and OS_3 is another objectset then $OS_1OS_3 \leq OS_2OS_3$ is always true. ☐

Theorem 2. *If an objectset is dominated by at least K other objectsets then we do not need to compose super objectsets for skyband-set computation with dominated objectset and all other objectsets except the dominant objectset.*

Proof: Assume OS_1, OS_2, OS_3, and OS_4 be the four objectsets in D. If objectset OS_1 is dominated by OS_2 and OS_3 then we do not need to compose super objectset OS_1OS_4 for $K = 2$. This is because according to Theorem 1 if $OS_2 \leq OS_1$ is true then for super objectset $OS_2OS_4 \leq OS_1OS_4$ is also true. Similarly, $OS_3OS_4 \leq OS_1OS_4$ is true. Therefore there exist at least two other objectset such as OS_2OS_4 and OS_3OS_4 that can dominate super objectset OS_1OS_4. Thus, it is proved that if an objectset is dominated by at least K other objectsets then we do not need to compose super objectsets for skyband-set computation with dominated objectset and all other objectsets except the dominant objectset. ☐

Theorem 1 and 2 gives us opportunity to obtain objectset dominance relationship without computing all objectsets. Table 1 shows that object O_4 is dominated by O_2 and O_3. By considering $\{O_1, O_5, O_6, O_7\}$ as non dominant objectset and using Theorem 2 without any computation as well as without any comparisons we get following dominance relationship for 2-objectsets $\{O_{1,2}, O_{1,3} \leq O_{1,4}\}, \{O_{2,5}, O_{3,5} \leq O_{4,5}\}, \{O_{2,6}, O_{3,6} \leq O_{4,6}\}, \{O_{2,7}, O_{3,7} \leq O_{4,7}\}$. Moreover, we also get two more dominance relationship for O_4 such as $O_{2,3} \leq O_{2,4}$ and $O_{2,3} \leq O_{3,4}$. Similarly, object O_5 is dominated by $\{O_2, O_6\}$ and using $\{O_1, O_3, O_4, O_7\}$ as non dominant objectsets give us dominance relationship $\{O_{1,2}, O_{1,6} \leq O_{1,5}\}, \{O_{2,3}, O_{3,6} \leq O_{3,5}\}, \{O_{2,4}, O_{4,6} \leq O_{4,5}\}, \{O_{2,7}, O_{6,7} \leq O_{5,7}\}$. In addition we also get $O_{2,6} \leq O_{2,5}$ and $O_{2,6} \leq O_{5,6}$. For $O_2 \leq O_6$ gives us $O_{1,2} \leq O_{1,6}, O_{2,3} \leq O_{3,6}, O_{2,4} \leq O_{4,6}, O_{2,5} \leq O_{5,6}$, and $O_{2,7} \not\leq O_{6,7}$. Finally, for $\{O_1, O_2, O_3, O_4, O_5, O_6 \leq O_7\}$ proposed method will compute following dominance relationship $\{O_{1,2}, O_{1,3}, O_{1,4}, O_{1,5}, O_{1,6} \leq O_{1,7}\}$, $\{O_{1,2}, O_{2,3}, O_{2,4}, O_{2,5}, O_{2,6} \leq O_{2,7}\}$, $\{O_{1,3}, O_{2,3}, O_{3,4}, O_{3,5}, O_{3,6} \leq O_{3,7}\}$, $\{O_{1,4}, O_{2,4}, O_{3,4}, O_{4,5}, O_{4,6} \leq O_{4,7}\}$, $\{O_{1,5}, O_{2,5}, O_{3,5}, O_{4,5}, O_{5,6} \leq O_{5,7}\}$, and $\{O_{1,6}, O_{2,6}, O_{3,6}, O_{4,6}, O_{5,6} \leq O_{6,7}\}$. Thus, according to Theorem 1 and Theorem 2 we can easily construct another *domRelationTable* for objectset size $s = 2$ using Table 1 without composing objectsets as well as performing any comparisons among them. The new *domRelationTable* is shown in Table 2.

Table 2. domRelationTable for 2-objectsets

Objectset	Dom. Objectset	Objectset	Dom. Objectset
$O_{1,2}$	$\varnothing$	$O_{3,4}$	$O_{2,3}$
$O_{1,3}$	$\varnothing$	$O_{3,5}$	$O_{2,3}, O_{3,6}$
$O_{1,4}$	$O_{1,2}, O_{1,3}$	$O_{3,6}$	$O_{2,3}$
$O_{1,5}$	$O_{1,2}, O_{1,6}$	$O_{3,7}$	$O_{1,3}, O_{2,3}, O_{3,4}, O_{3,5}, O_{3,6}$
$O_{1,6}$	$O_{1,2}$	$O_{4,5}$	$O_{2,5}, O_{3,5}, O_{2,4}, O_{4,6}$
$O_{1,7}$	$O_{1,2}, O_{1,3}, O_{1,4}, O_{1,5}, O_{1,6}$	$O_{4,6}$	$O_{2,6}, O_{3,6}, O_{2,4}$
$O_{2,3}$	$\varnothing$	$O_{4,7}$	$O_{1,4}, O_{2,4}, O_{3,4}, O_{4,5}, O_{4,6}, O_{2,7}, O_{3,7}$
$O_{2,4}$	$O_{2,3}$	$O_{5,6}$	$O_{2,6}, O_{2,5}$
$O_{2,5}$	$O_{2,6}$	$O_{5,7}$	$O_{1,5}, O_{2,5}, O_{3,5}, O_{4,5}, O_{5,6}, O_{2,7}, O_{6,7}$
$O_{2,6}$	$\varnothing$	$O_{6,7}$	$O_{1,6}, O_{2,6}, O_{3,6}, O_{4,6}, O_{5,6}, O_{2,7}$
$O_{2,7}$	$O_{1,2}, O_{2,3}, O_{2,4}, O_{2,5}, O_{2,6}$		

Dominance relation Table 2 retrieves candidate for objectsets skyband queries when objecetset size $s = 2$. For example if an user specify skyband-set query for $s = 2$ and $K = 1$, then from Table 2 proposed algorithm will retrieve candidate objectsets $\{O_{1,2}, O_{1,3}, O_{2,3}, O_{2,6}\}$. Therefore the proposed algorithm will compose only these four objectsets and perform domination check among them to obtain skyband-set result $\{O_{1,2}, O_{2,3}, O_{2,6}\}$. Here objectset $O_{1,3}$ is dominated by objectset $O_{2,6}$. Next, if the user is interested about skyband-set query with $s = 2$ and $K = 2$, then proposed algorithm will select candidate objectsets $\{O_{1,2}, O_{1,3}, O_{1,6}, O_{2,3}, O_{2,4}, O_{2,5}, O_{2,6}, O_{3,4}, O_{3,6}\}$ and perform domination check among these objectsets. Finally, it retrieves skyband-set query result as $\{O_{1,2}, O_{1,3}, O_{1,6}, O_{2,3}, O_{2,5}, O_{2,6}, O_{3,4}\}$. Therefore dominance relation Table 2 can retrieves candidate objectsets for any skyband-set query when $s = 2$. Proposed method will continue similar iterative procedure to construct dominance relation table each time for higher value of s and ready to answer skyband-set queries for any skyband value of K.

5   Performance Evaluation

We conducted a set of experiments with different dimensionalities (m), data cardinalities (n), and objectset size (s) to evaluate the effectiveness and efficiency of our proposed method. All experiments are run on a computer with Intel Core i7 CPU 3.4GHz and 4 GB main memory. We compiled the source codes under Java V6 in Windows 8.1 system. We also compared the performance with naive method. To make the comparison fair, we have include all the preprocessing cost, i.e., cost of naive method objectset generation. Each experiment is repeated five times and the average result is considered for performance evaluation. Three data distributions are considered as follows:

Correlated: a correlated dataset represents an environment in which, objects are good in one dimension are also good in the other dimensions. In a correlated dataset, fairly few objects dominate many other objects.

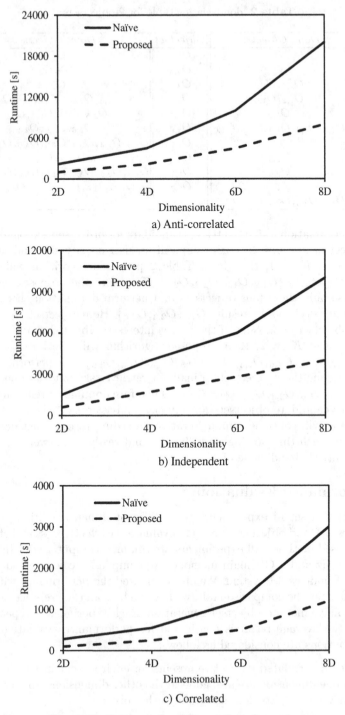

a) Anti-correlated

b) Independent

c) Correlated

Fig. 3. Performance for different data dimension

Anti-Correlated: an anti-correlated dataset represents an environment in which, if an object has small coordinates on some dimensions, it tends to have large coordinates on other dimensions or at least another dimension.

Independent: for this type of dataset, all attribute values are generated independently using uniform distribution. Under this distribution, the total number of non-dominating objects is between that of the correlated and the anti-correlated datasets.

5.1 Effect of Cardinality

For this experiment, we fix the data dimensionality m to 6, objectset size s to 3, and vary dataset cardinality n ranges from 2k to 10k. Figure 4(a), (b), and (c) shows the performance on correlated, independent, and anti-correlated datasets. Where both of the methods are highly affected by data cardinality. If the data cardinality increases then their performances decreases. The result shows that proposed method significantly outperforms the naive method. However, the performance of naive method degrades rapidly as the size of the dataset increases, especially when the data distribution is anti-correlated. This implies that proposed method can successfully avoid objectset composing as well as unnecessary comparisons.

5.2 Effect of Dimensionality

We study the effect of dimensionality on our technique. We fix the data cardinality n to 10k, objectset size s to 3 and vary dataset dimensionality m ranges from 2 to 8. The run-time results for this experiment are shown in Figure 3(a), (b), and (c). The result shows that as the dimension increases the performance of the both methods becomes slower. This is because for high dimension the number of non dominant objectset increases and the performance become slower. It is clear that the performance of the naive method is much worse than that of proposed method under all data distributions. However, the result on correlated data dataset is 10 times faster than independent data dataset. Where as it is 15 times faster than anti-correlated data dataset.

5.3 Effect of Objectset Size

In another experiment, we study the performance of proposed method under various objectset size s. We fix the data cardinality n to 10k and dataset dimensionality m to 6. The results are shown in Figure 5(a), (b), and (c). The result exhibits that as the objectset size s increases the performance of the both methods becomes slower. However, the result of naive method is much worse than that of proposed method when the value of objectset size s is greater than 1. The reason is in the beginning proposed method use SFS algorithm to construct $domRelationTable$ for $s = 1$, and performing domination check. After that for higher s it does not required to compose all objectset set as well as succeeded to avoid huge number of unnecessary comparisons.

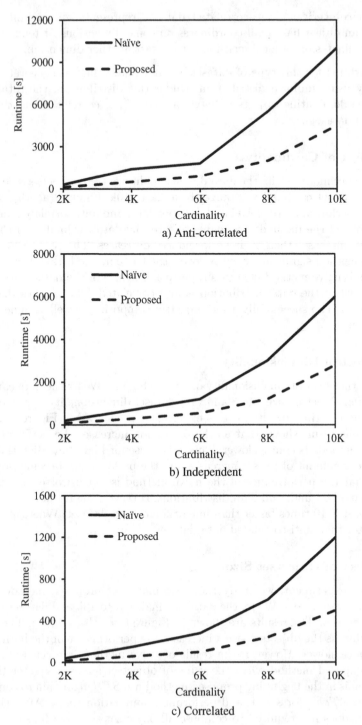

Fig. 4. Performance for different cardinality

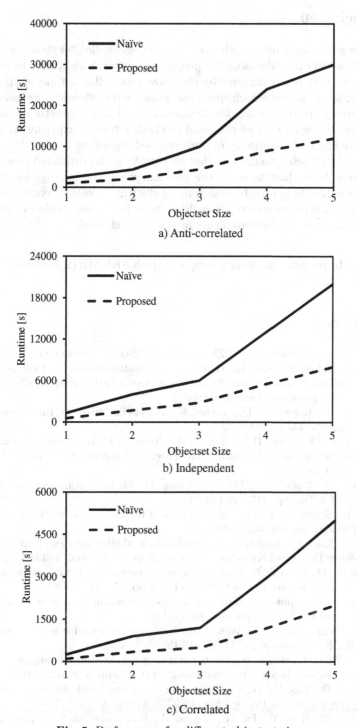

Fig. 5. Performance for different objectset size

6   Conclusion

This paper addresses a novel variant of skyline query called "skyband-set" query for sets of objects in a dataset. We propose an efficient algorithm to compute skyband-set query result for any objectset size s as well as for any skyband size K. In order to prune the search space and improve the efficiency, we have developed a pruning strategy using domRelationTable. Using synthetic datasets, we demonstrate the scalability of proposed method. Intensive experiments confirm the effectiveness and superiority of our proposed algorithm.

It is worthy of being mentioned that this work can be expanded in a number of directions. First, how to solve the problem when the aggregation function is not monotonic. Secondly, to design more efficient objectsets computation on distributed MapReduce architectures. Finally, to find small number of representative objectsets is another promising future research work.

Acknowledgments. This work is supported by KAKENHI (23500180, 25.03040) Japan.

References

1. Balke, W.-T., Güntzer, U., Zheng, J.X.: Efficient distributed skylining for web information systems. In: Bertino, E., Christodoulakis, S., Plexousakis, D., Christophides, V., Koubarakis, M., Böhm, K. (eds.) EDBT 2004. LNCS, vol. 2992, pp. 256–273. Springer, Heidelberg (2004)
2. Borzsonyi, S., Kossmann, D., Stocker, K.: The skyline operator. In: Proceedings of ICDE, pp. 421–430 (2001)
3. Chan, C.Y., Jagadish, H.V., Tan, K.-L., Tung, A.K.H., Zhang, Z.: Finding k-dominant skyline in high dimensional space. In: Proceedings of ACM SIGMOD, pp. 503–514 (2006)
4. Chomicki, J., Godfrey, P., Gryz, J., Liang, D.: Skyline with presorting. In: Proceedings of ICDE, pp. 717–719 (2003)
5. Dellis, E., Seeger, B.: Efficient computation of reverse skyline queries. In: Proceedings of VLDB, pp. 291–302 (2007)
6. Guo, X., Xiao, C., Ishikawa, Y.: Combination skyline queries. Transactions on Large-Scale Data- and Knowledge-Centered Systems VI 7600, 1–30 (2012)
7. Kossmann, D., Ramsak, F., Rost, S.: Shooting stars in the sky: An online algorithm for skyline queries. In: Proceedings of VLDB, pp. 275–286 (2002)
8. Lappas, T., Gunopulos, D.: Efficient confident search in large review corpora. In: Proc. PKDD Conference, pp. 467–478 (2010)
9. Lee, J., Hwang, S., Nie, Z., Wen, J.-R.: Navigation system for product search. In: Proc. ICDE Conference, pp. 1113–1116 (2010)
10. Lin, X., Yuan, Y., Wang, W., Lu, H.: Stabbing the sky: Efficient skyline computation over sliding windows. In: Proceedings of ICDE, pp. 502–513 (2005)
11. Papadias, D., Tao, Y., Fu, G., Seeger, B.: An optimal and progressive algorithm for skyline queries. In: Proceedings of SIGMOD, pp. 467–478 (2003)
12. Papadias, D., Tao, Y., Fu, G., Seeger, B.: Progressive skyline computation in database systems. ACM Transactions on Database Systems 30(1), 41–82 (2005)

13. Roy, S.B., Amer-Yahia, S., Chawla, A., Das, G., Yu, C.: Constructing and exploring composite items. In: Proceedings of SIGMOD, pp. 843–854 (2010)
14. Siddique, M.A., Morimoto, Y.: Algorithm for computing convex skyline object-sets on numerical databases. IEICE Transactions on Information and Systems E93-D(10), 2709–2716 (2010)
15. Su, I.-F., Chung, Y.-C., Lee, C.: Top-k combinatorial skyline queries. In: Proceedings of DASFAA, pp. 79–93 (2010)
16. Tan, K.-L., Eng, P.-K., Ooi, B.C.: Efficient progressive skyline computation. In: Proceedings of VLDB, pp. 301–310 (2001)
17. Tao, Y., Xiao, X., Pei, J.: Subsky: Efficient computation of skylines in subspaces. In: Proceedings of ICDE, pp. 65–65 (2006)
18. Vlachou, A., Doulkeridis, C., Kotidis, Y., Vazirgiannis, M.: Skypeer: Efficient subspace skyline computation over distributed data. In: Proceedings of ICDE, pp. 416–425 (2007)
19. Wan, Q., Wong, R.C.-W., Peng, Y.: Finding top-k profitable products. In: Proceedings of ICDE, pp. 1055–1066 (2011)
20. Wang, G., Xin, J., Chen, L., Liu, Y.: Energy efficient reverse skyline query processing over wireless sensor networks. IEEE Transactions Knowledge Data Engineering (TKDE) 24(7), 1259–1275 (2012)
21. Xia, T., Zhang, D., Tao, Y.: On skylining with flexible dominance relation. In: Proceedings of ICDE, pp. 1397–1399 (2008)
22. Zou, L., Chen, L., Ozsu, M.T., Zhao, D.: Dynamic skyline queries in large graphs. In: Proc. DASFAA Conference, pp. 62–78 (2010)

Towards an Ontology-Based
Generic Pipeline Editor

Paolo Bottoni and Miguel Ceriani

Sapienza, University of Rome, Italy
{bottoni,ceriani}@di.uniroma1.it

Abstract. The pipeline concept is widely used in computer science to represent non-sequential computations, from scientific workflows to streaming transformation languages. While pipelines stand out as a highly visual representation of computation, several pipeline languages lack visual editors of production quality. We propose a method by which a generic pipeline editor can be built, centralizing the features needed to maintain and edit different pipeline languages. To foster adoption, especially in less programming-savvy communities, the proposed visual editor will be web-based. An ontology-based approach is adopted for the description of both the general features of the pipelines and the specific languages to be supported. Concepts, properties and constraints are defined using the Web Ontology Language (OWL), providing grounding in existing standards and extensibility. The work also leverages existing ontologies defined for scientific worlkflows.

1 Introduction

The concept of pipeline of processes dates back to the 60's, when Douglas McIlroy was working on UNIX shells and devised a mechanism to compose data handling commands executed in parallel. The concept is now widely used in computer science to represent a form of non-sequential computation. Pipelines visualize the dataflow paradigm, where a sub-process is run anytime its inputs are available.

They are used in the context of the description of workflows, where processes are either instances of a library of local operators or remote calls to web services. They are especially well suited when a top down approach is mostly effective, like for scientific workflows in which researchers often need to reuse workflows defined by other researchers outside their own group or even their own community.

Pipelines are also used to represent transformations or actions in a way that permits streaming evaluation. Examples range from UNIX system pipes to languages offering complex manipulation capabilities of structured data, like XProc for XML documents [1]. The pipeline metaphor can also be used for declarative data languages in which a *data view* is built on top of data sources or other data views through a set of basic operators, e.g. relational algebra operators. Data languages of this kind have been used in several contexts: database management systems, complex event processing, ETL systems, Semantic Web applications.

W. Chu et al. (Eds.): DNIS 2015, LNCS 8999, pp. 56–73, 2015.

Pipelines stand out as a highly visual representation of computation, so it is natural to expect to use visual editors for creation and modification of pipelines. Moreover, visual programming helps significantly in reducing barriers to programming in communities where programming skills are unusual.

Despite this, visual pipeline editors do not keep up with the available pipeline languages. Some pipeline languages have no visual editor, while others have visual editors that remained of prototypical quality or that are not maintained. To develop a visual editor for new or existing pipeline languages is a non-trivial effort, thus hindering the adoption potential of this kind of languages.

In this work we explore the possibility of building a generic visual pipeline editor. In the proposed vision the editor centralizes the features needed to maintain and edit pipelines. Each specific pipeline language is defined through a profile that specifies the available components, their properties, the constraints and how the pipeline is converted to the native language. To foster adoption, especially in less programming-savvy communities, the proposed visual editor is Web-based.

While the editor does not need to be aware of the full semantics of each specific language, it needs some information on the syntax, expressed somehow in the language profile. We propose an ontology-based approach, in which components, properties and constraints of each specific language are defined using the Web Ontology Language (OWL), an established standard offering great expressiveness to define and constrain specific languages. The basic pipeline concepts – common to all the languages – are defined in an ontology too. This is the base ontology from which the ontologies representing specific languages will extend.

OWL is defined on top of the Resource Description Framework (RDF) data model. The standard query language of RDF (SPARQL) can thus be used to map a pipeline defined through a specific ontology to the corresponding native pipeline language.

In the rest of the paper, after discussing technology background in Sect. 2 and related work in Sect. 3, Sect. 4 describes the proposed method. Sect. 5 presents a proof-of-concept implementation inside an existing project. Finally, Sect. 6 discusses conclusions and future work.

2 Scientific/Technological Context

In this chapter the context of the work will be sketched, briefly introducing the relevant scientific and technological concepts.

2.1 The Semantic Web

In 1999 Tim Berners-Lee coined the term Semantic Web to describe a future Web of machine-processable data. The vision encompassed the extension of the World Wide Web from a web of documents to a web of data and a way to attach actionable semantics to the data, to allow autonomous agents to infer useful facts [2]. While some ideas proposed by Berners-Lee and colleagues are still unrealized, in the last few years a lot of structured data became available and started being reused on the Web. This happened in part thanks to a new concept introduced by Berners-Lee in 2006: *Linked Data* [3]. Linked Data is used to refer to a set of recipes to make the web of data work in practice. Today the terms Semantic Web and Linked Data are used somehow interchangeably, the second one putting less focus on the use of logical models to formally define the semantics of data. In the rest of the section the Semantic Web technologies used in this work are discussed.

Resource Description Framework. The relational model is widely used to represent virtually any kind of structured information. The Resource Description Framework (RDF) [4] generalises the relational model to the universe of structured data in the World Wide Web. In the RDF data model, knowledge is represented via *RDF statements* about *resources*, where a resource is an abstraction of any piece of information about some domain. An RDF statement is represented by an *RDF triple*, composed of *subject* (a resource), *predicate* (specified by a resource as well) and *object* (a resource or a literal, i.e. a value from a basic type). An *RDF graph* is therefore a set of RDF triples. Resources are uniquely identified by a Uniform Resource Identifier (URI) [5], or by a local (to the RDF graph) identifier if they have no meaning outside of the local context (in which case they are called *blank nodes*). The resources used to specify predicates are called *properties*. A resource may have one or more *types*, specified by the predefined property rdf:type. An *RDF dataset* is a set of graphs, each associated with a different name (a URI), plus a default graph without a name. We use RDF through the framework to represent any kind of information and its transformations. In RDF, *prefixes* can be used in place of the initial part of a URI, representing specific namespaces for vocabularies or sets of resources.

Different syntaxes have been defined for writing RDF. In the present work the Turtle syntax [6] will be mostly used for its conciseness and compatibility with parts of SPARQL syntax (described in 2.1). The system is nevertheless designed to be compatible with all the established RDF syntaxes.

Vocabularies. One key requirement to provide reusability on linked data is to reuse existing vocabularies (especially if widely used as general purpose vocabularies or established in the specific relevant community). An RDF vocabulary is a set of classes and properties (and possibly specific resources) to be used with a certain meaning. The maintainer of a vocabulary is encouraged to publish it in a machine readable format, documenting formally the restrictions to its use, together with the intended meaning expressed in natural language.

The RDF Schema [7] is a vocabulary used in turn for defining vocabularies. It can be used to define the class hierarchy, possibly a property hierarchy and define domain and range of properties. To define more complex constraints – e.g., that a class has a certain property – a more powerful language is used: the Ontology Web Language (OWL) [8].

SPARQL. We extensively use **SPARQL**[1] [9], the standard query language for RDF datasets. SPARQL has a relational algebra semantics, analogous to those

---

[1] Originally a recursive acronym SPARQL Protocol and RDF Query Language, the extended form has then been dropped from W3C documents.

of traditional relational languages, such as Structured Query Language (SQL). The current version of the standard is SPARQL 1.1, but much of the existing literature refers to the previous version, SPARQL 1.0 [10]. The SPARQL 1.1 algebra offers an expanded set of operators, effectively allowing the expression of queries that were not expressible before.

A SPARQL query takes as input an RDF dataset and can be in one of four different forms: SELECT, CONSTRUCT, ASK or DESCRIBE. In the present work only the first two forms are used. The basic building block of SPARQL is the *triple pattern*, a triple in which each of the components can be replaced by a variable. A *basic graph pattern* is a set of triple patterns associated with a specific input graph (the default graph, a named graph specified by an URI or a generic named graph specified by a variable). Each basic graph pattern is matched against the input dataset and the result is a relation in which each tuple corresponds to a variable set assignment. The relations generated through basic graph patterns are composed using relational operators. The output SPARQL SELECT query is, as for a SQL query, an ordered multiset of tuples.

The SPARQL CONSTRUCT, on the contrary, produces as output an RDF graph. A CONSTRUCT query uses another kind of element called *triple template*, similar to *triple pattern* but using only the variables exposed in the top level relation. A set of triple templates is thus used to build the output graph, adding for each tuple of the relation the triples obtained by substituting the variables in the triple templates.

While the SPARQL Query Language is "read-only", the SPARQL Update Language [11] defines a way to perform updates on a *Graph Store*, the "modifiable" version of an RDF Dataset. A SPARQL Update *request* is composed of a number of *operations*. There are different available kinds of operations, corresponding to different data manipulation needs. In this work the DELETE/INSERT operation is used, in which the modification of a dataset is specified through a query from which the sets of deleted and added triples are derived.

2.2 Rich Web Applications

The ubiquity of Web browsers and Web document formats across a range of platforms and devices drives developers to build applications on the Web and its standards. Requirements for browsers have dramatically changed since the first days of the Web. Now a browser is an interface to an ever-growing set of client capabilities, exemplified by the **Rich Web Client** activity at W3C[2]. All modern browsers support the Scalable Vector Graphics (SVG) standard [12], a language representing mixed vector and raster content, based on Extensible Markup Language (XML) [13]. Together with the long established Document Object Model (DOM) events [14,15] and ECMAScript[3] [16] support, SVG allows the realisation of complete interactive visualisation applications. Indeed, ECMAScript libraries for interactive data visualization are proliferating, from

---

[2] http://www.w3.org/2006/rwc/Activity.html
[3] Commonly called JavaScript, the dialect from Mozilla Foundation.

standard visualisations [17,18,19] to specialized ones for specific domains [20], especially leveraging the SVG technology.

2.3 Pipeline and Workflow Systems

During the last decade workflow systems have been gaining traction. They offer complex functionality based on the integration of distributed web services. While the initial model of workflow systems was mainly procedural and based on XML syntax and tools, recently there is a trend in adding semantic information to different components of the Web services and the workflow engines.

Semantic annotation of Web services typically adds some semantics to a syntax descriptions based on XML – e.g. the Web Based Description Language (WSDL) [21]. While WSDL gives information about the expected input and output structure (described in turn using XML Schema [22,23]), RDF-based semantic annotation vocabularies are used to describe the purpose of the service and some properties of its implementation – e.g. security considerations. A number of vocabularies have been proposed for the purpose – of which OWL-S [24] is especially noteworthy for its comprehensive approach – and in 2007 W3C produced the recommendation Semantic Annotations for WSDL and XML Schema (SAWSDL) [25]. However none of the proposed vocabularies gained widespread adoption, possibly due to some inherent complexity not balanced by immediate advantages (partly due to lack of support from production tools).

Recently, other methods for semantical descriptions of Web services have been proposed, avoiding the XML-based approach but leveraging the REST paradigm [26] and providing a stronger connection between services input/output and RDF representation. Examples are ReLL [27], RESTdesc [28], SSWAP [29] and SADI [30]. A specific effort is SPARQL 1.1 Service Description [31], describing properties and features of SPARQL endpoint or graph store. It is still to be seen if these vocabularies will get more traction than WSDL-based ones. The purpose of the semantic information associated with the Web services is to allow some (centralized or decentralized) agents to (semi-)automatically look for services and possibly even choose between them at runtime.

A complementary effort (coinciding with the previous one for some comprehensive languages as OWL-S), is to use semantics in the workflows. These have been introduced especially for scientific workflows, in three main roles:

- to describe and/or annotate the workflow itself, e.g. for reproducibility and discussion on a scientific workflow [32,33];
- to describe the executable behaviour of the workflow, e.g., declaratively defining service composition [24,34];
- to describe and/or annotate the data itself, e.g. for keeping provenance information [35];
- to define data transformations, e.g. rule-based transformations on data [36].

3   Related Work

A big amount of research work has been dedicated to different types of pipeline languages, especially to workflow systems. In 2.3 some of the proposals for modeling workflow systems have been cited. A smaller, but interesting, amount of effort has gone to the study of the visual and interactive aspects of workflows and pipelines (see for example [37,38]). The work presented here is in a way orthogonal to these efforts, in the sense that it combines the modeling of the pipeline language with the modeling of its visualization/interaction.

For the authors it has been motivating the work carried on by Gesing et al. [39]. They recognize the need for "Consistent Web-Based Workflow Editors" supporting multiple workflow languages and propose thus an application based on an entity-relationship model representing the available languages and their components.

The present work gets partial inspiration from their model but differ from their design mainly for being based on ontologies rather than on an entity-relationship model. OWL ontologies permit to define more complex constraints on the languages syntax, simplifies the independent development of different language profiles, and allow the system to be potentially distributed.

4   Proposed Method

In this section, the method proposed to build the generic editor is described. At first an abstract model is proposed, then it is given a more concrete facet by proposing specific technologies and ontologies.

4.1   The Model

The system, as we envision it, is composed of the following components:

- an ontology able to represent generic pipeline components and their connections;
- a Web-based generic pipeline editor driven by specific *language profiles*;
- a mechanism for import/export from/to the target language format, also based on *language profiles*.

A *language profile* is the way a specific pipeline language is specified for the system. Each *language profile* consists of the following components:

- an ontology that represents the specific components and constraints of the language, specializing the concepts defined in the generic ontology;
- some code snippets to be plugged in the editor to customize specific behavior;
- a declarative specification of the transformations from/to the target language –in a way understood by the generic system.

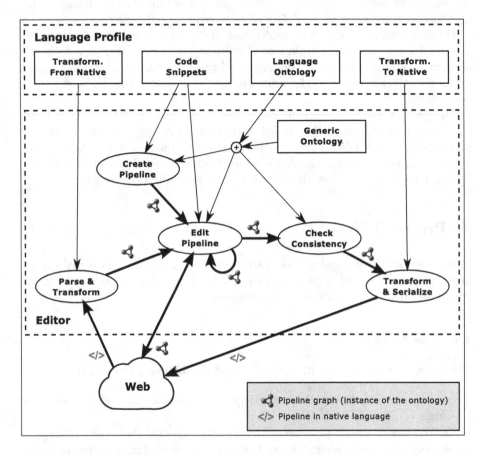

Fig. 1. Model proposed for building the generic pipeline editor

In Figure 1, the proposed model is shown.

The specific ontology may specify constraints on the syntax of the pipeline, from the type of parameters and inputs/outputs, to complex constraints on the structure. User interface components can make use of the provided constraints to guide the user in building the pipeline and configuring the components. A formal validation is done in a separate step, previous to export, to ensure that all the constraints are fulfilled.

In the editor a pipeline is represented by a graph structure, instance of the language ontology –merged with the generic ontology. It should be noted that the graph structure may differ from the pipeline native code in a number of ways:

- the graph has visual information used for editing that cannot be encoded in the native language;
- correspondence between visual components and native language is not necessary one-to-one;
- the format may be different.

The composition of the transformations *from native* and *to native* should in any case lead to a semantically equivalent pipeline.

4.2   Technologies

The model proposed in 4.1 must be realised through a suitable set of technology choices. As the editor is designed to be Web-based and scalable through the Web infrastructure, we propose to use a set of established Web standards –anticipated in 2.1 and 2.2.

More in detail, the following technologies are proposed:

- the Web Ontology Language (OWL) to define both the generic ontology and each specific language ontology;
- the RDF query language SPARQL to declaratively define the transformations from/to the native language;
- an RDF based representation for the native pipeline language format –e.g., an RDF representation of XML, if the pipeline language is XML-based;
- a rich web application for the actual creation, editing and management of the pipelines, based on HTML, Scalable Vector Graphics (SVG), and JavaScript technologies;
- language specific code snippets in JavaScript, to customize creation, parsing, and serialization of each component.

In Figure 2 the model is shown again, showing now in detail the proposed technologies. *Format profiles* have been included in the model to represent the basic format upon which the syntax of the specific language is built: e.g., XML, JSON. As SPARQL is used to define the transformations to and from the native language, the language must be represented in the RDF model. The format profile provides hence an ontology through which the basic format (e.g., XML)

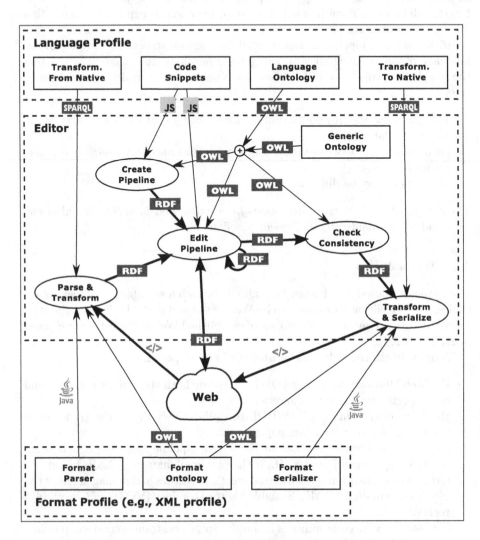

Fig. 2. The technologies used for building the generic pipeline editor

can be represented in RDF and the implementations of a parser and a serializer to convert between the native format and its RDF-based representation. Parser and serializer are implemented in Java. Pipeline languages defined on top of the RDF model do not need a format profile. The XML format profile is specially useful as most used pipeline languages are based on XML.

4.3   Ontology

While providing a formal definition of the concept of *pipeline*, is outside the scope of this paper, a generic description must be attempted to frame the present proposal. A pipeline – for the purposes of this paper – consists of a set of components (also called processes) from a finite set of component types interconnected by connectors (also called channels), thus forming a graph (that depending on the language may have topology constraints, like acyclicity). Each component may have some inputs and/or some outputs and the connectors are defined as going from an output of a component to an input of another ome –or even to the same component, if this is allowed. A pipeline, on its own, has a set of inputs – called sources – and a set of outputs –called sinks. Pipelines can be often used in turn as components – directly or through an appropriate component type – to compose new pipelines based on existing ones. It is important to stress that this informal definition is *on purpose* minimal and agnostic with respect to the exact pipeline semantics: the goal is to model the minimal structure needed by a visual editor.

In order to represent this concept of generic pipelines a new ontology could be designed. Nevertheless, it is worth building on the considerable amount of work that has been done in modeling contiguous or overlapping domains. In particular, much research is aimed at the construction of ontologies representing and/or annotating workflows. As the concepts and relationships required to represent pipelines overlap with several existing workflow ontologies, we looked for the ones that do not add unnecessary complexity to what required by the proposed pipeline editor and are actively maintained. At least two ontologies designed for scientific workflows satisfy the requirements:

- mecomp[4] in the myExperiment ontology [32];
- wfdesc[5] in the Workflow for Ever (Wf4Ever) Research Object Model [33].

The myExperiment project [40] is an on-going effort to provide a Web portal and technology framework to facilitate and encourage researchers to share their workflows and methods in a re-usable way. The myExperiment ontology [32] was designed for the RDF representation of scientific workflows and their relationships with other research products – e.g. researchers' personal data, papers, data sets. Having being designed by the developers of the Taverna workflow system [41], the workflow model follows the structure of Taverna workflow language.

---

[4] http://rdf.myexperiment.org/ontologies/components/
[5] http://purl.org/wf4ever/wfdesc#

The Wf4Ever Research Object Model has been recently proposed for the same purpose, but not tied to a particular portal or workflow system. Concerning the description of scientific workflows the two ontologies have very few differences. In the present work, the myExperiment ontology is used,as it is generic enough for our purposes and more established than the Wf4Ever model. Nevertheless, due to the similarity of the two models, the proposed pipeline editor would work also with the Wf4Ever model with small changes.

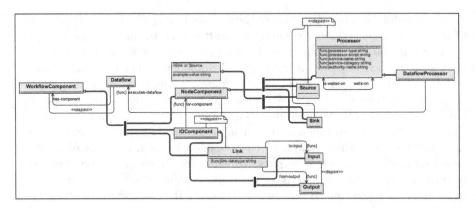

Fig. 3. The mecomp fragment of myExperiment ontology

The mecomp fragment of myExperiment ontology –shown in Figure 3 through the compact representation proposed by Bārzdiņš et al. [42]– is used specifically to describe the structure of scientific workflows and it will be used here to describe the general structure of pipelines. The ontology defines a `mecomp:Dataflow` as composed by a set of resources of type `mecomp:WorkflowComponent`. The latter can be a `mecomp:NodeComponent` (corresponding to one of the components of the pipeline), a `mecomp:IOComponent` (an input or output gate on a component) or a `mecomp:Link` (the connection from an output gate to an input gate). A `mecomp:NodeComponent` can in turn be a `mecomp:Source` (an input of the pipeline), a `mecomp:Sink` (an output of the pipeline) or `mecomp:Processor` (any other component). The myExperiment ontology reuses terms of the Dublin Core ontology to provide metadata properties: textual descriptions and identifiers for the dataflow components. The myExperiment ontology, however, does not provide a way to express visual features like the layout and appearance of the components in the editor. For that purpose, terms of the Visualization Ontology (VISO) [43] have been used. VISO has been developed to classify the multiple facets of data visualization applications, but in this case only terms from the GRAPHIC module (graphic relations and representations) are used: `graphic:x_position` and `graphic:y_position` to specify the visual positioning of a component; `graphic:shape_named` and `graphic:color_named` to specify shape and color of an input gate of a component (for output gates the appearance is fixed in the current version). VISO is work in progress and some of these properties are still considered *not stable*, nevertheless it appeared to be worth reusing ontology terms in the process of being consolidated, compared with developing new terms.

A specific pipeline language ontology must extend myExperiment ontology providing specific components and constraints. At the very least it has to define a set of subclasses of `mecomp:Processor` class. Other classes may be subclassed as well, e.g. to define different kinds of link.

5   Implementation

A proof-of-concept implementation of the presented method was realized as part of SWOWS, an on-going effort to build a platform for declarative linked data applications [44,45]. In SWOWS, applications are built as pipelines of RDF operators.

The existing SWOWS pipeline editor has been redesigned to provide support for generic pipeline languages, following the method described in Section 4. In the rest of this section the main components of the application are described.

5.1   The Repository

Figure 4 shows the structure of the pipeline repository, based on the Callimachus Web server [46], a Content Management System (CMS) based on linked data.

Callimachus contains internally an instance of an OpenRDF Sesame triple store, maintaining the data that represents the structure of the content. The actual content of each uploaded file is stored in blobs (in the case of RDF files they are stored both as blobs and as triples). Any content that is managed by Callimachus is an instance of one of a set specific OWL classes such as *Page, Folder, File, User*[6].

The Web interface of Callimachus exposes the content of a server instance in three ways:

- Web pages for view, creation, editing of resources associated via templates to each class;
- REST APIs for programmatic access to resources, also dependent on their class;
- direct SPARQL access to the underlying triple store.

Callimachus applications are composed of instances of existing classes, possibly new classes, and supporting resources and code. For the pipeline repository, a Callimachus application has been developed: it contains the generic ontology and the common user interface components. Specific language profiles may be included in the same application package – e.g., a set of predefined languages – or as separate packages –possibly developed independently, as long as the same generic ontology is extended and the same user interface components are reused.

---

[6] The full ontology is at `http://callimachusproject.org/rdf/2009/framework`

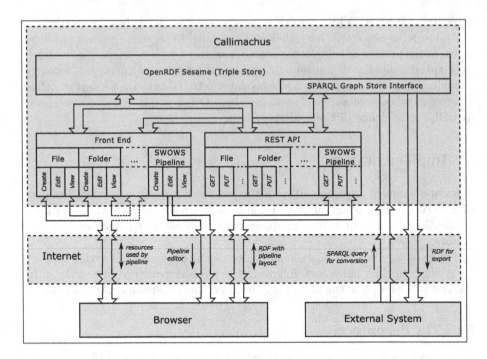

Fig. 4. The blocks composing the repository and its interplay with a browser

A pipeline is represented in Callimachus as an RDF graph using the corresponding specific ontology. The visual pipeline editor is associated in Callimachus with the activity of editing a resource of type mecomp:Dataflow. As mecomp:Dataflow represents a generic pipeline, the association is actually used for the concrete subclasses of mecomp:Dataflow, corresponding to the specific languages. In this way some parts of the user interface – whose implementation will be described more in detail in 5.2 – can be customized for the specific language.

When a pipeline starts being edited, first the visual editor and then the RDF graph representing the pipeline are downloaded to the client. The RDF graph is converted by the editor in visual form to be visualized, modified by the user, and saved back to the server as RDF graph at user request. The modules for execution of the conversions to and from the native language format of the pipeline are not included in the currently described implementation. In the special case in which the basic format of the native language is RDF, only one (or a set of) SPARQL query(ies) for the export and one (or a set) for the import have to be executed. In this case, for the export the query(ies) can be executed by Callimachus obtaining directly the pipeline in native language. As for the import, the query(ies) can be sent by Callimachus to the endpoint holding the imported pipeline, obtaining the pipeline as instance of the derived ontology.

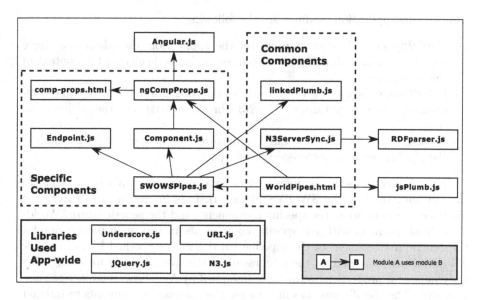

Fig. 5. The modules (HTML and Javascript) composing the editor

5.2 The Editor

The editor is implemented as a rich Web application with its client side logic coded combining HTML, CSS, JavaScript and embedded in the Callimachus Web application described in Section 5. The main modules and their relationship are shown in Figure 5.

The JavaScript application has been built on top of the following libraries:

- *Underscore* for functional programming operators in JavaScript [47];
- *jQuery* to abstract from browsers' idiosyncrasies [48];
- *URI* for URIs manipulation;
- *N3* for RDF in-memory manipulation and parsing (serializing) from (to) Turtle syntax [49];
- *RDFparser* to parse RDF/XML syntax [50];
- *AngularJS* for form templates with two way data-binding [51];
- *jsPlumb* to visually manage the connections between components [52].

The last two libraries are central in building the user interface in a flexible way. AngularJS permits the declarative definition of an HTML-based piece of user interface, in such a way that the user interface is separate from the underlying model and the synchronization between the two is automated. In the app AngularJS is used as the base to flexibly define forms for the configuration of specific component properties.

jsPlumb is a library to visually show and manipulate pipeline structures. Using jsPlumb most of the low level work for visual interaction with the pipeline is avoided. A jsPlumb instance manages a set of endpoints (the inputs/outputs of the components) and the connections between them.

The main application modules are the following:

- *WorldPipes.html*, the entry point for the application; the editor is a single page application, all the other resources are hence loaded in the context of this web page;
- *linkedPlumb*, the module that interfaces the RDF graph of the pipeline (represented in-memory through the N3 library format) and the jsPlumb instance;
- *N3ServerSync*, the module that synchronizes the in-memory RDF graph of the pipeline with its server-side counterpart.

Figure 5 also shows the modules specific to a language (in this example DfPL, the language used in the SWOWS platform). It is interesting to analyze the interface points between the specific components and the generic editor. World-Pipes.html interacts with the specific components in two main points: one for the generic management of the pipeline (in this case embodied by the module *SWOWSPipes*) and one for the integration of the specific forms for configuring the component properties (in this case embodied by AngularJS module *ngComp-Props.js*). The specific modules in turn use the common components to interact with the server (through N3ServerSync) and with the jsPlumb instance (through linkedPlumb).

6   Conclusions and Future Work

The authors recognized the need for a common generic pipeline editor and proposed a general method to build such a system, centered around an extensible ontological representation of pipelines.

The feasibility of this approach has been shown through the use of existing established technologies. Furthermore, a proof-of-concept implementation has been realized in a specific context.

The implementation is not still a full fledged generic pipeline editor. To that purpose a more formal analysis of the interfaces between the generic editor and the specific components is needed.

Nevertheless, the authors hope above all that with this work may contribute to stimulate a discussion on the topic and on the ways to to tackle this issue.

References

1. Walsh, N., Milowski, A., Thompson, H.: XProc: An XML Pipeline Language. W3C REC (May 11, 2010)
2. Berners-Lee, T., Hendler, J., Lassila, O.: The Semantic Web. Scientific American 284(5), 34–43 (2001)
3. Berners-Lee, T.: Linked data Design Issues (2006), http://www.w3.org/DesignIssues/LinkedData.html
4. Cyganiak, R., Wood, D., Lanthaler, M.: RDF 1.1 Concepts and Abstract Syntax. W3C REC (February 25, 2014)

5. Berners-Lee, T., Fielding, R., Masinter, L.: Uniform Resource Identifier (URI): Generic Syntax. RFC 3986 (INTERNET STANDARD) Updated by RFC 6874 (January 2005)
6. Beckett, D., Berners-Lee, T., Prud'hommeaux, E., Carothers, G.: RDF 1.1 Turtle: Terse RDF Triple Language. W3C REC (February 25, 2014)
7. Brickley, D., Guha, R.: RDF Schema 1.1. W3C REC (February 25, 2014)
8. Motik, B., Patel-Schneider, P.F., Parsia, B., Bock, C., Fokoue, A., Haase, P., Hoekstra, R., Horrocks, I., Ruttenberg, A., Sattler, U., Smith, M.: OWL 2 Web Ontology Language: Structural Specification and Functional-Style Syntax, 2nd edn. W3C REC 11 (December 2012)
9. Harris, S., et al.: SPARQL 1.1 Query Language. W3C REC (March 21, 2013)
10. Prud'hommeaux, E., Seaborne, A.: SPARQL Query Language for RDF. W3C REC (January 15, 2008)
11. Schenk, S., Gearon, P., et al.: SPARQL 1.1 Update. W3C REC (March 21, 2013)
12. Dahlström, E., Dengler, P., Grasso, A., Lilley, C., McCormack, C., Schepers, D., Watt, J., Ferraiolo, J., Fujisawa, J., Jackson, D.: Scalable Vector Graphics (SVG) 1.1. 2nd edn. W3C REC (August 16, 2011)
13. Bray, T., Paoli, J., Sperberg-McQueen, C.M., Maler, E., Yergeau, F., Cowan, J.: Extensible Markup Language (XML) 1.1. 2nd edn. W3C REC, edited in place (September 29, 2006) (August 16, 2006)
14. Pixley, T.: Document object model (DOM) level 2 events specification. W3C REC (November 13, 2000)
15. Kacmarcik, G., Leithead, T., Rossi, J., Schepers, D., Höhrmann, B., Le Hégaret, P., Pixley, T.: Document Object Model (DOM) Level 3 Events Specification. W3C REC (November 13, 2000)
16. ECMA: ECMAScript Language Specification, Standard ECMA-262, 5.1 edn (2011), http://www.ecma-international.org/ecma-262/5.1/
17. Google: Google charts (2010)
18. Belmonte, N.G.: JavaScript InfoVis Toolkit (2011), http://philogb.github.io/jit/
19. Bostock, M., Ogievetsky, V., Heer, J.: D3: Data-driven documents. IEEE Trans. Visualization & Comp. Graphics 17, 2301–2309 (2011)
20. Smits, S.A., Ouverney, C.C.: jsPhyloSVG: A Javascript Library for Visualizing Interactive and Vector-Based Phylogenetic Trees on the Web. PloS One 5(8), e12267 (2010)
21. Chinnici, R., Canon, J.J.M., Ryman, A., WeerawaranaBeckett, S.: Web Services Description Language (WSDL) Version 2.0 Part 1: Core Language. W3C REC (June 26, 2007)
22. Gao, S.S., Sperberg-McQueen, C.M., Thompson, H.S.: W3C XML Schema Definition Language (XSD) 1.1 Part 1: Structures. W3C REC (April 5, 2012)
23. Peterson, D., Gao, S.S., Malhotra, A., Sperberg-McQueen, C.M., Thompson, H.S.: W3C XML Schema Definition Language (XSD) 1.1 Part 2: Datatypes. W3C REC (April 5, 2012)
24. Martin, D.: OWL-S: Semantic Markup for Web Services. W3C Member Submission (November 22, 2004)
25. Farrell, J., Lausen, H.: Semantic Annotations for WSDL and XML Schema. W3C REC (August 28, 2007)
26. Fielding, R.T., Taylor, R.N.: Principled design of the modern Web architecture. ACM Transactions on Internet Technology (TOIT) 2(2), 115–150 (2002)
27. Alarcon, R., Wilde, E.: Linking data from restful services. In: Third Workshop on Linked Data on the Web, Raleigh, North Carolina (April 2010)

28. Verborgh, R., Steiner, T., Van Deursen, D., De Roo, J., Van de Walle, R., Vallés, J.G.: Capturing the functionality of Web services with functional descriptions. Multimedia Tools and Applications 64(2), 365–387 (2013)

29. Gessler, D.D., Schiltz, G.S., May, G.D., Avraham, S., Town, C.D., Grant, D., Nelson, R.T.: Sswap: A simple semantic web architecture and protocol for semantic web services. BMC Bioinformatics 10(1), 309 (2009)

30. Wilkinson, M.D., Vandervalk, B.P., McCarthy, E.L., et al.: The Semantic Automated Discovery and Integration (SADI) Web service Design-Pattern, API and Reference Implementation. J. Biomedical Semantics 2, 8 (2011)

31. Williams, G.T.: SPARQL 1.1 Service Description. W3C REC (March 21, 2013)

32. Newman, D., Bechhofer, S., De Roure, D.: myExperiment: An ontology for e-Research. In: Proc. SWASD at ISWC-2009, vol. 523. CEUR-WS (2009)

33. Belhajjame, K., Zhao, J., Garijo, D., Hettne, K., Palma, R., Corcho, Ó., Gómez-Pérez, J.M., Bechhofer, S., Klyne, G., Goble, C.: The research object suite of ontologies: Sharing and exchanging research data and methods on the open web. arXiv preprint arXiv:1401.4307 (2014)

34. Weigand, H., van den Heuvel, W.J., Hiel, M.: Rule-based service composition and service-oriented business rule management. In: Proceedings of the International Workshop on Regulations Modelling and Deployment, ReMoD 2008, Citeseer, pp. 1–12 (2008)

35. Lebo, T., Sahoo, S., McGuinness, D.: PROV-O: The PROV Ontology. W3C REC (April 30, 2013)

36. Coppens, S., Verborgh, R., Mannens, E., Van de Walle, R.: Self-sustaining platforms: A semantic workflow engine. In: COLD (2013)

37. Wood, J., Wright, H., Brodie, K.: Collaborative visualization. In: Proceedings of the IEEE Visualization 1997, pp. 253–259 (1997)

38. Maguire, E., Rocca-Serra, P., Sansone, S.A., Davies, J., Chen, M.: Taxonomy-based glyph design&# 8212; with a case study on visualizing workflows of biological experiments. IEEE Transactions on Visualization and Computer Graphics 18(12), 2603–2612 (2012)

39. Gesing, S., Atkinson, M., Klampanos, I., Galea, M., Berthold, M.R., Barbera, R., Scardaci, D., Terstyanszky, G., Kiss, T., Kacsuk, P.: The demand for consistent web-based workflow editors. In: Proceedings of the 8th Workshop on Workflows in Support of Large-Scale Science, pp. 112–123. ACM (2013)

40. Goble, C.A., Bhagat, J., Aleksejevs, S., Cruickshank, D., Michaelides, D., Newman, D., Borkum, M., Bechhofer, S., Roos, M., Li, P., et al.: myexperiment: A repository and social network for the sharing of bioinformatics workflows. Nucleic Acids Research 38(suppl. 2), W677–W682 (2010)

41. Oinn, T., Addis, M., Ferris, J., Marvin, D., Senger, M., Greenwood, M., Carver, T., Glover, K., Pocock, M.R., Wipat, A., et al.: Taverna: A tool for the composition and enactment of bioinformatics workflows. Bioinformatics 20(17), 3045–3054 (2004)

42. Bārzdiņš, J., Bārzdiņš, G., Čerāns, K., Liepiņš, R., Sprogis, A.: UML Style Graphical Notation and Editor for OWL 2. In: Forbrig, P., Günther, H. (eds.) BIR 2010. LNBIP, vol. 64, pp. 102–114. Springer, Heidelberg (2010)

43. Polowinski, J., Voigt, M.: Viso: A shared, formal knowledge base as a foundation for semi-automatic infovis systems. In: Proc. CHI 2013, pp. 1791–1796. ACM (2013)

44. Bottoni, P., Ceriani, M.: A Dataflow Platform for In-silico Experiments Based on Linked Data. In: Madaan, A., Kikuchi, S., Bhalla, S. (eds.) DNIS 2014. LNCS, vol. 8381, pp. 112–131. Springer, Heidelberg (2014)

45. Bottoni, P., Ceriani, M.: SWOWS and dynamic queries to build browsing applications on linked data. J. Vis. Lang. Comput. 25(6), 738–744 (2014)

46. Battle, S., Wood, D., Leigh, J., Ruth, L.: The Callimachus Project: RDFa as a Web Template Language. In: Proc. COLD (2012)
47. Fogus, M.: Functional JavaScript: Introducing Functional Programming with Underscore. js. OʼReilly Media, Inc. (2013)
48. Bibeault, B., Katz, Y.: jQuery in Action. Manning Publications Co., Greenwich (2008)
49. Verborgh, R.: N3.js: Lightning fast, asynchronous, streaming Turtle for JavaScript (2012), https://github.com/RubenVerborgh/N3.js
50. Ley, J.: Simple javascript RDF Parser and query thingy (2006), http://jibbering.com/rdf-parser/
51. Darwin, P.B., Kozlowski, P.: AngularJS web application development. Packt Publishing (2013)
52. Porritt, S.: jsPlumb JavaScript library: Visual connectivity for webapps (2011), http://www.jsplumb.org/

Synthetic Evidential Study
as Primordial Soup of Conversation

Toyoaki Nishida[1,*], Atsushi Nakazawa[1], Yoshimasa Ohmoto[1], Christian Nitschke[1],
Yasser Mohammad[2], Sutasinee Thovuttikul[1], Divesh Lala[1],
Masakazu Abe[1], and Takashi Ookaki[1]

[1] Graduate School of Informatics, Kyoto University, Sakyo-ku, Kyoto, Japan
{nishida,nakazawa.atsushi,
ohmoto,christian.nitschke}@i.kyoto-u.ac.jp,
{thovutti,lala,abe,ookaki}@ii.ist.i.kyoto-u.ac.jp
[2] Assiut University, Assiut, Egypt
yasserfarouk@gmail.com

Abstract. Synthetic evidential study (SES for short) is a novel technology-enhanced methodology for combining theatrical role play and group discussion to help people spin stories by bringing together partial thoughts and evidences. SES not only serves as a methodology for authoring stories and games but also exploits the framework of game framework to help people sustain in-depth learning. In this paper, we present the conceptual framework of SES, a computational platform that supports the SES workshops, and advanced technologies for increasing the utility of SES. The SES is currently under development. We discuss conceptual issues and technical details to delineate how much we can implement the idea with our technology and how much challenges are left for the future work.

Keywords: Inside understanding, group discussion and learning, intelligent virtual agents, theatrical role play, narrative technology.

1   Introduction

Our world is filled with mysteries, ranging from fictions to science and history. Mysteries bring about plenty of curiosities that motivate creative discussions, raising interesting questions, such as "how did Romeo die in Romeo and Juliet?", "why was Julius Caesar assassinated?", and "why did Dr. Shuji Nakamura, 2014 Nobel laureate, leave Japan to become a US citizen?" to name just a few.

It is quite natural that mysteries motivate people to bring together the partial arguments, ranging from thoughts to evidences or theories, to derive a consistent and coherent interpretation of the given "mystery". Let us call this discussion style an *evidential study* if it places a certain degree of emphasis on logical consistency. Discussions need not be deductive so long as they are based on a discipline such as an abduction. On the one hand, rigorous objective discussions were considered mandatory in an academic

---

* Corresponding author.

W. Chu et al. (Eds.): DNIS 2015, LNCS 8999, pp. 74–83, 2015.
© Springer International Publishing Switzerland 2015

disciplines such as archeology and evidential methods broadly observed in empirical sciences. On the other hand, the constraints are much weaker in the folk activities such as those workshops in schools. Even though they may sometimes run into flaw, their utility in education, e.g., motivating students, is even higher due to its playful aspects.

Unfortunately, "naive" evidential study has its own limitations as well even with enhancement by conventional IT. Although normal groupware greatly helps large scale online discussions and accumulate their results for later use, it does not permit the participants to examine the discussion from many perspectives, which is critical in evidential study. In general, conventional software falls short for allowing participants to visualize their thoughts instantly on the spot for sharing better understanding. In particular, evidential study cannot be deepened without going deeply into the mental process of the characters in question by simulating how they perceive the world from their first person view, it is critical to uncover the mental process of a given character by deeply into partial interpretation in particular.

Synthetic evidential study (SES for short) is a novel technology-enhanced methodology for combining theatrical role play and group discussion to help people spin stories by bringing together partial thoughts and evidences [1]. SES practically implements the idea of primordial soup of conversation in conversational informatics [2], in which common ground and communicative intelligence may co-evolve through the accumulation of conversations. SES leverages the powerful game technologies [3], and is applicable to not just mysteries but also a wide range of application in science and technology, e.g., [4-6].

2   The Conceptual Framework of SES

SES combines theatrical role play and group discussion to help people spin stories by bringing together partial thoughts and evidences. The conceptual framework of SES is shown in Fig. 1.

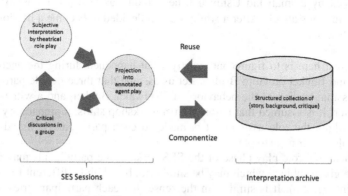

Fig. 1. The conceptual framework of SES

At the top level, the SES framework consists of the SES sessions and the interpretation archives. In each SES session, participants repeat a cycle of a theatrical role play, its projection into an annotated agent play, and a group discussion. One or more successive execution of SES sessions until participants come to a (temporary) satisfaction is called a SES workshop.

In the theatrical role play phase, participants play respective roles to demonstrate their first-person interpretation in a virtual space. It allows them to interpret the given subject from the viewpoint of an assigned role.

In the projection phase, an annotated agent play on a game engine is produced from the theatrical role play in the previous phase by applying the oral edit commands (if any) to theatrical actions by actors elicited from the all behaviors of actors. We employ annotated agent play for reuse, refinement, and extension in the later SES sessions.

In the critical discussion phase, the participants or other audience share the third-person interpretation played by the actors for criticism. The actors revise the virtual play until they are satisfied. The understanding of the given theme will be progressively deepened by repeatedly looking at embodied interpretation from the first- and third- person views.

The interpretation archive logistically supports the SES sessions. The annotated agent plays and stories resulting from SES workshops may be decomposed into components for later reuse so that participants in subsequent SES workshops can adapt previous annotated agent plays and stories to use as a part of the present annotated agent play.

Let us use a hypothetical example to illustrate on more details. Consider the following story is given as a subject for a SES workshops:

> The Ushiwaka-Benkei story: When Ushiwaka, a young successor to a noble Samurai family who once was influential but killed by the opponents, walked out of a temple in a mountain in the suburbs of Kyoto where he was confined, to wander around the city as daily practice, he met Benkei, a strong priest Samurai on Gojo Bridge. Although Benkei tried to punish him as a result of having been provoked by a small kid Ushiwaka, he couldn't as Ushiwaka was so smart to avoid Benkei's attack. After a while, Benkei decided to become a life-long guard for Ushiwaka.

One interest here is to figure out exactly what happened during the encounter of Ushiwaka and Benkei on Gojo Bridge. Let us assume that three people participate in the SES session, to discuss the behaviors of Ushiwaka, Benkei, and a witness during the encounter. It is assumed that before a SES workshop starts, a preparatory meeting is held in which the role assignment is made so each participant is asked to think about the role she or he is to play.

In the theatrical role play phase of the SES session, the participants may do a role play for the given subject, which may be similar to, but slightly different from the one as shown in Fig. 2(a). It is similar, in the sense that each participant tries to bodily express her/his interpretation. It is different, as participants are expected play in a shared virtual space in the SES theatrical role play phase and we do not assume that participants need to physically play together, need carry props such as a sword prop, or physically jump from a physical stage, though production of a virtual image and motion is a challenge.

Fig. 2. (a) Left: Theatrical role play by the SES participants is similar to, but slightly different from the role play like this. (b) Right: Agent play to be reproduced from theatrical role play on the left. One or more parts of the agent play may be annotated with the player's comments.

A think aloud method is employed so each actor can not only show her/his interpretation but also show the rationale for the interpretation, describe their intention of each action using an oral editing command, or even criticize role play by other actors. The actor's utterances will be recorded and used as a resource for annotations associated with the agent play. It is an excellent opportunity for each participant to gain a pseudo-experience of the situation through the angle of the given role's viewpoint. Each actor's behaviors are recorded using audio-visual means.

The goal of the projection phase is to reproduce an annotated agent play such as shown in Fig. 2(b) from the theatrical role play.

Ideally, the projection should be automatically executed on-line while participants are acting in the theatrical role play phase. In order to do so, separation of the behaviors of each actor into the genuine theatrical role play, the editing behavior, and the meta-level actions such as commentating.

In the critical discussion phase, the participants watch the resulting annotated agent play and discuss the resulting interpretation from various angles (Fig. 3).

Fig. 3. Critical discussion based on the annotated agent play will help participants to deepen objective understanding. In actual SES sessions, the participants may well meet virtually to criticize the annotated agent play.

Again, we do not necessarily assume that the participants need to hold a physical meeting. Most importantly, the participants are provided with an opportunity to discuss the theatrical role play from the objective third person view, in addition to the subjective view gained in the theatrical role play phase. A 3D game engine such as Unity allows the group to switch between the objective and subjective views of a given role to deepen the discussion. Furthermore, it might be quite useful if the participants can modify the annotated agent play on the fly during the discussions, while we assume that it will be more convenient to make major revisions in the theatrical role play phase. That is why we have assumed that the SES sessions will be repeated.

3   Computational Platform for SES

We are building the SES computational platform by combining a game engine (Unity 3D) and the ICIE+DEAL technology we developed for conversational informatics research [2]. ICIE is an immersive interaction environment made available by a 360-degree display and surround speakers and audio-visual sensors for measuring the user's behaviors. ICIE consists of an immersive audio-visual display and plug-in sensors that capture user behaviors therein. The user receives an immersive audio-visual information display in a space surrounded by 8 large displays about 2.5 m diameter and 2.5 m in height (Fig. 4). We have implemented a motion capture consisting of multiple depth sensors that can sense the user's body motion in this narrow space. This platform constitutes a "cell" that will allow for the user to interact with each other in a highly-immersive interface. Cells can be connected with each other or other kinds of interfaces such as a robot so that the users can participate in interactions in a shared space.

Fig. 4. The hardware configuration of ICIE

DEAL is a software platform that permits individual software components to cooperate to provide various composite services for the interoperating ICIE cells. Each server has one or more clients that read/writes information on a shared blackboard on the server. Servers can be interconnected on the Internet and the blackboard data can be shared. One can extend a client with plug-ins using DLL. Alternatively a client can join the DEAL platform by using DLL as a normal application that bundles network protocols for connecting a server.

By leveraging the immersive display presentation system operated on the DEAL platform and the motion capture system for a narrow space it allows for projecting the behaviors of a human to those of an animated character who habits in a shared virtual space. The ICIE+DEAL platform is coupled with the Unity (http://unity3d.com/) platform so the participants can work together in a distributed environment. It permits the participants to animated Unity characters in the virtual space.

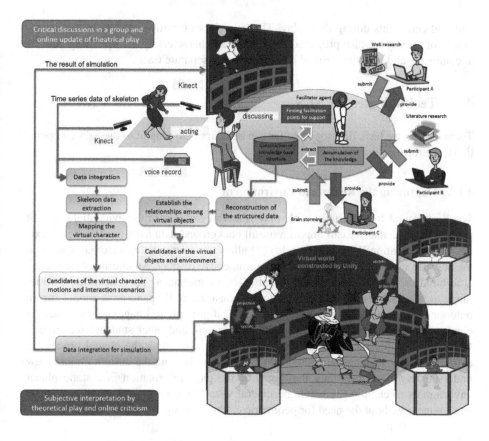

Fig. 5. The architecture of the SES computational platform

Fig. 5 shows how ICIE and DEAL can be put together to support theatrical role play and group discussion in SES. The upper half of the diagram depicts a subsystem for supporting group discussions. It consists of a group discussion support for conventional evidential study based on literature and review of theatrical role play. A facilitation agent is employed to help participants bring together partial knowledge and criticisms to formulate a consistent flow of theatrical role plays as a synthetic interpretation of the subject. The major role of the facilitation agent is to estimate and track tacit preferences of participants underlying discussions and navigate the discussion by conducting appropriate facilitation behaviors such as encouraging remark by a participant with a conflicting opinion at a proper timing. In addition, the subsystem permits the participants to revise the details of the behavior of the theatrical role play on the fly by using a mini-studio adjacent to the discussion table.

The lower half of Fig. 5 depicts a subsystem for supporting a theatrical role play in a distributed environment. ICIEs can be interconnected on the net to allow participants from remote sites to share a virtual space for interaction to collaboratively coordinate a theatrical role play to build up their subjective interpretation of the subject. Think aloud method is used to permit the participants to make critical or even editorial comments during their play. The behaviors of participants are not only recorded for review but also projected onto synthetic characters by incorporating editorial comments and isolating critical comments into a separate track.

4   Technologies for SES

Technologies we have developed so far can be adapted to plug in to the SES computational platform.

4.1   Capturing the Realworld Environment

The background for a theatrical role play is an integral part of the evidential study. It is often the case that we can import a useful background data from the Internet. Flexible Communication World (FCWorld) [7] allows us to project an external source on the net such as Google Street View to the immersive display. An alternative approach is to build the data for the shared virtual space by measuring the realworld. A wide-range area visualizer [8] can automatically construct a 3D panoramic model for an outdoor scene from a collection of ordinary 2D digital photo images for the place by combining structure from motion, multi view stereo, and other standard techniques. Capturing first person view and gaze is critical in estimating the mental process of actor. Our corneal imaging method allows for going beyond the current human view understandings that uses only the 'point' of the gaze information in a static planar environment to capture the whole peripheral visual field in an arbitrary dynamic 3D environment without the need for geometric calibration [9].

4.2   Capturing Realworld Interaction

Capturing the motion of an actor in an immersive interaction cell becomes ready with existing technologies, by mapping the skeleton data from a Kinect to animate a Unity character. However, we need to overcome several limitations.

A theatrical role play by multiple actors can be captured by our 3DCCbyMK technology [2]. It exploits Kinect technology to measure and critique physical display of interpretation played together one or more local participants. The system allows us to simultaneously capture the behaviors of up to four participants, by integrating data captured by multiple Kinects, thereby to project the interaction by multiple participants to the shared virtual space as it is. It assumes that the scene consists of the static background and moving people in the foreground.

Two subsystems were developed for capturing the static background and the dynamic foreground. OpenNI is used to capture an RGB image, depth map, skeleton data, user masks and tracking IDs. To reconstruct a static 3D model for the surrounding background, a single Kinect is carried around in the environment to gather data from continuous viewpoints. SLAM (simultaneous localization and mapping) is used to build a local 3D model from the data from each point. Image features calculated by the SURF method [10] are used to calculate the similarity to integrate the local 3D model into the global 3D model. The LMedS method [11] is used to reduce errors in image feature matching.

The motion estimation subsystem of the 3DCCbyMK technology estimates the motion of each participant using the skeleton data from multiple Kinect sensors. Time series data are checked to reduce the confusion between the left and right joints. The output of the two subsystems are integrated to produce an interaction scene. A problem remains regarding incorporating the results into the 3D coordinate system of a game engine.

4.3   Estimating Attitudes and Emotion

Mind reading or estimating the internal mental state of the human from external cues or social signals is necessary to build an advanced service for the SES participants, such as discussion support as discussed later in this section or even producing a better annotated agent play in the theatrical role play.

DEEP [12] is a method for estimating a dynamically changing emphasizing point by integrating verbal behaviors, body movements, and physiological signals of the user. DEEP is applied to situations where many known/unknown factors must be considered in choosing a satisfactory option. DEEP repeats the cycle consisting of explanation, demand-seeking, and completion-check until the user is satisfied with the proposed option. User's emphasizing point is obtained at the same time. gDEEP is a method for estimating emphasizing point for a group using DEEP as a component. gDEEP repeats the cycle similar to DEEP.

4.4   Discussion Support

Our discussion support method centers on a chairperson agent that can support discussions by estimating distribution of opinions, engagement, and emphasizing points. We addressed interactive decision-making during which people dynamically and interactively change the emphasizing points and have built several prototypes of a group decision-making support agent [12-14].

A recent prototype [14] can guide the divergence and convergence processes [15] in the facilitation by producing appropriate social signals as a result of grasping the status of the decision-making process by the group of participants. The system uses the gDEEP method to estimate the emphasizing point of the group from verbal presentation of demands and nonverbal and physiological reactions to the information presentation by the agent. If it has turned out that the group has not yet well formulated an emphasizing point, the system will present information obtained from a broad search in the problem space with reference to the emphasizing point so that it can stimulate the group's interest to encourage divergent thought. When it has turned out that the group has formulated an emphasizing point, the system will focus on the details to help the group carry out convergent thought for making decision.

Our technologies allow for capturing not only explicit social signals that clearly manifest on the surface but also tacit and ambiguous cues by integrating audio-visual and physiological sensing.

4.5   Fluid Imitation Learning

Learning by mimicking is a computational framework for producing the interactive behaviors of conversational agents from a corpus obtained from the WOZ experiment. In this framework, the learning robot initially "watches" how people interact with each other, estimates communication principles underlying the way the target actor communicates with other partners, and applies the estimated communication patterns to produce the communicative behaviors of the conversation agent.

Currently, we focus on nonverbal behaviors and approximate the communicative behaviors as a collection of continuous time series.   A suite of unsupervised learning algorithms [16,17] are used to realize the idea.

By having this algorithm in combination with action segmentation and motif discovery algorithms that we developed, we have a complete fluid imitation engine that allows the robot to decide for itself what to imitate and actually carry on the imitation [18].

We designed, implemented and evaluated of a closed loop pose copying system [19]. This system allows the robot to copy a single pose without any knowledge of velocity/acceleration information and using only closed loop mathematical formulae that are general enough to be applicable to most available humanoid robots. This system makes it possible to reliably teach a humanoid by demonstration without the need of difficult to perform kinesthetic teaching.

5   Concluding Remark

Synthetic evidential study (SES) combines theatrical role play and group discussion to help people spin stories. The SES framework consists of (a) the SES sessions of theatrical role playing, projection into the annotated agent plays and a critical group discussions, and (b) a supporting interpretation archive containing annotated agent plays and stories. We have described the conceptual framework of SES, a computational platform that supports the SES workshops, and advanced technologies for increasing the utility of SES. The SES is currently under development. So far, we have implemented basic components. The system integration and evaluation are left for future work. Future challenges include, among others, automatic production of actor's intended play, self-organization of the interpretation archive, and automatic generation of the virtual audience from critical discussions.

Acknowledgment. This study has been carried out with financial support from the Center of Innovation Program from JST, JSPS KAKENHI Grant Number 24240023, and AFOSR/AOARD Grant No. FA2386-14-1-0005.

References

1. Nishida, T., et al.: Synthetic Evidential Study as Augmented Collective Thought Process – – Preliminary Report. ACIIDS 2015, Bali, Indonesia (to be presented, 2015)
2. Nishida, T., Nakazawa, A., Ohmoto, Y., Mohammad, Y.: Conversational Informatics: A Data-Intensive Approach with Emphasis on Nonverbal Communication. Springer (2014)

3. Harrigan, P., Wardrip-Fruin, N.: Second Person: Role-Playing and Story in Games and Playable Media, MIT Press (2007)
4. Thovuttikul, S., Lala, D., van Kleef, N., Ohmoto, Y., Nishida, T.: Comparing People's Preference on Culture-Dependent Queuing Behaviors in a Simulated Crowd, In: Proc. ICCI*CC 2012, pp. 153–162 (2012)
5. Lala, D., Mohammad, Y., Nishida, T.: A joint activity theory analysis of body interactions in multiplayer virtual basketball. Presented at 28th British Human Computer Interaction Conference, September 9-12, Southport, UK (2014)
6. Tatsumi, S., Mohammad, Y., Ohmoto, Y., Nishida, T.: Detection of Hidden Laughter for Human-agen Interaction. Procedia Computer Science (35), 1053–1062 (2014)
7. Lala, D., Nitschke, C., Nishida, T.: Enhancing Communication through Distributed Mixed Reality. AMT 2014, 501–512 (2014)
8. Mori, S., Ohmoto, Y., Nishida, T.: Constructing immersive virtual space for HAI with photos. In: Proceedings of the IEEE International Conference on Granular Computing, pp. 479–484 (2011)
9. Nitschke, C., Nakazawa, A., Nishida, T.: I See What You See: Point of Gaze Estimation from Corneal Images. In: Proc. 2nd IAPR Asian Conference on Pattern Recognition (ACPR), pp. 298–304 (2013)
10. Bay, H., Tuytelaars, T., Van Gool, L.: Surf: speeded up robust features. In: Leonardis, A., Bischof, H., Pinz, A. (eds.) ECCV 2006, Part I. LNCS, vol. 3951, pp. 404–417. Springer, Heidelberg (2006)
11. Zhang, Z.: Parameter estimation techniques: A tutorial with application to conic fitting. Image Vis. Comput. 15(1), 59–76 (1997)
12. Ohmoto, Y., Miyake, T., Nishida, T.: Dynamic estimation of emphasizing points for user satisfaction evaluations. In: Proc. the 34th Annual Conference of the Cognitive Science Society, pp. 2115–2120 (2012)
13. Ohmoto, Y., Kataoka, M., Nishida, T.: Extended methods to dynamically estimate emphasizing points for group decision-making and their evaluation. Procedia-Social and Behavioral Sciences 97, 147–155 (2013)
14. Ohmoto, Y., Kataoka, M., Nishida, T.: The effect of convergent interaction using subjective opinions in the decision-making process. In: Proc. the 36th Annual Conference of the Cognitive Science Society, pp. 2711–2716 (2014)
15. Kaner, S.: Facilitator's guide to participatory decision-making. Wiley (2007)
16. Mohammad, Y., Nishida, T., Okada, S.: Unsupervised simultaneous learning of gestures, actions and their associations for Human-Robot Interaction. In: IROS 2009, pp. 2537–2544 (2009)
17. Mohammad, Y., Nishida, T.: Learning interaction protocols using Augmented Bayesian Networks applied to guided navigation. In: IROS 2010, pp. 4119–4126 (2010)
18. Mohammad, Y., Nishida, T.: Robust Learning from Demonstrations using Multidimensional SAX. Presented at ICCAS 2014, October 22-25, Korea (2014)
19. Mohammad, Y., Nishida, T.: Tackling the Correspondence Problem - Closed-Form Solution for Gesture Imitation by a Humanoid's Upper Body. In: Yoshida, T., Kou, G., Skowron, A., Cao, J., Hacid, H., Zhong, N. (eds.) AMT 2013. LNCS, vol. 8210, pp. 84–95. Springer, Heidelberg (2013)

Understanding Software Provisioning:
An Ontological View

Evgeny Pyshkin[1], Andrey Kuznetsov[2], and Vitaly Klyuev[3]

[1] St. Petersburg State Polytechnic University
Institute of Computing and Control
21, Polytekhnicheskaya st., St. Petersburg 195251, Russia
`pyshkin@icc.spbstu.ru`
[2] Motorola Solutions, Inc.
St. Petersburg Software Center
Business Centre T4, 12, Sedova st., St. Petersburg 192019, Russia
`andrei.kuznetsov@motorolasolutions.com`
[3] University of Aizu
Software Engineering Lab.
Tsuruga, Ikki-Machi, Aizu-Wakamatsu 965-8580, Japan
`vkluev@u-aizu.ac.jp`

Abstract. In the areas involving data relatedness analysis and big data processing (such as information retrieval and data mining) one of common ways to test developed algorithms is to deal with their software implementations. Deploying software as services is one of possible ways to support better access to research algorithms, test collections and third party components as well as their easier distribution. While provisioning software to computing clouds researchers often face difficulties in process of software deployment. Most research software programs utilize different types of unified interface; among them there are many desktop command-line console applications which are unsuitable for execution in networked or distributed environments. This significantly complicates the process of distributing research software via computing clouds. As a part of knowledge driven approach to provisioning CLI software in clouds we introduce a novel subject domain ontology which is purposed to describe and support processes of software building, configuration and execution. We pay special attention to the process of fixing recoverable build and execution errors automatically. We study how ontologies targeting specific build and runtime environments can be defined by using the software provisioning ontology as a conceptual core. We examine how the proposed ontology can be used in order to define knowledge base rules for an expert system controlling the process of provisioning applications to computing clouds and making them accessible as web services.

Keywords: Knowledge Engineering, Ontology Design, Service-Oriented Architecture, Cloud Computing, Software Deployment Automation.

W. Chu et al. (Eds.): DNIS 2015, LNCS 8999, pp. 84–111, 2015.

1 Introduction

Service-oriented software and cloud computing significantly transformed the ways we think about possibilities to provide access to numerous computational resources. Currently there are many efforts about investigating possibilities to use clouds in organizing research and collaborative work. Particularly, in the domain of music information retrieval (MIR) researchers often develop software programs implementing different algorithms. Many of such implementations still remain desktop applications with command-line interface (CLI applications). Originally they are not intended to be executed in networked or distributed environments. With respect to issues of facilitating access to such implementations we have to mention the following problems:

1. A client local machine (where an application is assumed to be executed) might not well suit long lasting resource consuming computations.
2. A local application is hardly accessible for other researchers that wish to use the algorithm.
3. Client algorithms might require access to huge test collections which are hard to download and to allocate on a desktop storage.
4. Some test collections (particularly, in music information retrieval (MIR)) might not be publicly available due to the copyright restrictions.

Research communities often complain that many software implementations remain unpublished. Hence, it is often difficult to compare results achieved by different researchers even if we have access to the test data they used. There is another challenge: one researcher is often unable to reproduce results of others (even if they are published).

In the domain of MIR the MIREX[1] was organized to improve algorithms distribution and evaluation with using wide range of evaluation tasks [1]. However, since the evaluation process is implemented in the form of an annual competition researchers are unable to experiment with MIREX tests at any time they need. Researchers have very limited access to the solutions and test collections of others.

First, let us take a look at the Vamp system[2] [2]. Basically, the Vamp provides a plugin based framework which makes the published MIR research software accessible by the people outside the developer field. Properly speaking, the Vamp defines an application programming interface forcing developers to unify the application interface which has to conform the framework requirements. Despite the Vamp became a popular mean of MIR algorithms distribution, the MIREX still accepts CLI software implementations. Hence, the problem of CLI software provisioning remains to be of current concern.

---

[1] MIREX – Music Information Retrieval Evaluation eXchange:
   http://www.music-ir.org/mirex/
[2] http://www.vamp-plugins.org

Second, we have to mention the NEMA environment[3] [3]. The NEMA creates a distributed networked infrastructure providing access to the MIR software and data sets over the Internet. The NEMA architecture supports publishing client algorithms as services implementing the required web interface. The deployment process is partially automated: it is based on using a set of preconfigured virtual machine images. Each image provides a platform (e.g. *Python, Java*, etc.) which is completely configured to be used by the NEMA flow service. Image modifications are manual, hence developers are expected to have specific knowledge in order to deploy their solutions.

The MEDEA[4] system [4] enables deploying an arbitrary CLI application on an arbitrary cloud platform. Special attention is paid to such cloud services as *Google App Engine, EC2, Eucalyptus*, and *Microsoft Azure*. The MEDEA uses standard virtual machine images provided by a cloud and uploads a special wrapper (a task worker, in MEDEA terms) to the running virtual machine. The wrapper initializes the respective runtime environment (e.g. *Python, Java*, etc.) and executes a client application. Unfortunately, execution and configuration errors are not analyzed, there is no any automatic mechanism to handle build, execution and configuration errors conditioned not by the client code bugs, but by the deployment process faults.

What are advantages we expect in provisioning research software in clouds? First, clients are able to get quicker access to remote resources, be it a network storage, a virtual machine or an applied software. Second, server side computing decreases requirements for the client side hardware. Third, a public API can be defined for a deployed cloud application: publishing the application source code is not required, neither distributing software binary code. Fourth, test collections stored within a cloud are not necessary to be downloaded to client machines. In turn, clients require no direct access to test collections: there are less problems conditioned by copyright restrictions. Furthermore, clients pay only for the resources rent for short periods of time. Finally, application deployment and installation issues have to be resolved only once.

Despite there are many research efforts aimed to facilitate using cloud services for provisioning research software, there are still many open questions. Let's make closer inspection of the problem. There are applications developed for clouds and there are tools automating the deployment process. However, CLI applications are commonly not intended to be executed in clouds, hence they require different automation tools. Existing tools usually proceed from the assumption that a *virtual platform* (e.g. a PaaS or an IaaS service)[5] is configured properly. It means that all external dependencies (e.g. libraries, external resources, compiler or library versions, etc.) are resolved, and the only problem is to upload and to run the code in a cloud [4]. As a matter of actual practice,

---

[3] NEMA – Networked Environment for Music Analysis:
www.music-ir.org/?q-nema/overview
[4] MEDEA – Message, Enqueue, Dequeue, Execute, Access.
[5] We refer both a PaaS (platform as a service) and an IaaS (infrastructure as a service) computing node as a *virtual platform* if the difference between them is not important.

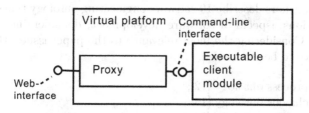

Fig. 1. A web-interface proxy is required in order to deploy CLI applications as web services

this assumption is rarely true, and deploying becomes very complicated: research software developers (being experts in their subject domain) might not be experienced enough in programming and in system administration. They might have no special knowledge on how to transfer their applications to a cloud, and how to configure a cloud runtime environment so as the application works the same way as it works on a local machine.

A reasonable way to deliver a CLI application to a cloud is to define a wrapper (or a proxy) providing a web interface to the application as Figure 1 shows. An automation tool implements mechanisms of transferring a client module (coupled with the related data) to a cloud service. The question is how to discover and to handle software execution errors *automatically*, with special emphasis on the errors conditioned by possible misconfiguration of a virtual platform. In [5] we introduced an idea to use a knowledge driven approach based on an ontology purposed to support processes of software building, its configuration and execution and to describe build and execution errors and actions required to fix recoverable errors automatically. In this work we describe this approach in greater details with a particular focus on ontology design.

2 Introducing the Ontology Usage Domain

As is the case of software testing, there are two major approaches to discover and handle configuration errors and client code external dependencies: source and object modules *static analysis* and *dynamic analysis* (of the execution with using some test data). Static analysis is used in such methods and algorithms as *ConfAnalyzer* [6], *ConfDiagnoser* [7] and *ConfDebugger* [8]. Well-known examples of dynamic analysis tools are *AutoBash* [9] and *ConfAid* [10].

2.1 Sources

In order to formalize error description and the relationships between an error and its resolution procedure, knowledge engineering formalisms are required. Despite there isn't any well-known ontology describing processes of software build, run and runtime environment configuration, we have to cite some general-purpose ontologies used in software engineering.

In [11] the authors describe 19 software engineering ontology types. They give a coat to various aspects of software development, documenting, testing and maintenance. Considering the most relevance to the paper issues, the following ontologies have to be mentioned:

- Software process ontology [12];
- System behavior ontology [13];
- Software artifact ontology [14];
- System configuration ontology [15].

Each of the above listed formalisms operates with a subset of concepts related to the software build, run and environment configuration, but none of these covers the subject in a whole. Beyond subject domain ontologies there are well-known meta-ontologies which potentially fit a wide variety of problems. The process ontologies (like *PSL* [16], *BPEL* [17], *BPMN* [18]) or the general purpose ontology (like *Cyc* [19]) can serve as examples of such formalisms. Unfortunately, meta-ontologies do not contain any concept or relationship which would allow us to describe a specific knowledge area and its specific problems.

In the following sections we define an ontology aimed to describe and resolve problems of CLI software provisioning to a virtual platform. We examine how to construct a production knowledge base used to support client application automatic deployment in clouds. Let us mention again that in this paper we are focused on deployment problems in relation to the special software class – *research software implementing algorithms developed in MIR mostly*. The authors of such programs are usually able to implement an algorithm in the form of CLI console application which transforms input data to the output according to the data formats required by a certain algorithm evaluation system. However it is common that they might not be experienced enough to resolve runtime environment failures or to guarantee that virtual platform requirements are satisfied.

2.2   Target Architecture

It seems impossible and probably useless to try to define an ontology "in the vacuum". In this paper we briefly examine the architectural aspects closely related to the ontology definition. We follow our work [20] where we introduced a target architecture consisting of a virtual platform, a cloud broker and a deployment manager (see Figure 2), the latter being responsible for successful deployment of CLI client console applications on a virtual platform. However that paper doesn't contain any detailed explanation of the software provisioning ontology.

A knowledge base (KB), an inference engine and its working memory are components of an expert system controlling the provisioning process. A pre-installed deployment manager agent provides a web interface to a client application, gathers information about the client application state and environment state. The deployment manager agent uses the knowledge base to resolve deployment problems. The agent interacts with the configuration manager by using high-level commands which describe required actions in order to reconfigure the platform.

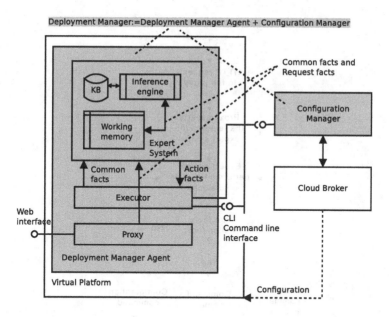

Fig. 2. Managing application deployment in the cloud: an architectural view

In turn, the configuration manager interacts with the cloud broker by using low-level commands. In so doing, it controls the installation of required cartridges[6] as well as the recreation of virtual machines if necessary. Figure 3 represents the general deploying procedure where the expert system is used to resolve possible configuration errors.

There are two major ways to upload a client module on a virtual platform:

– An executable module is uploaded by a client.
– An executable module is generated as a result of a server side building process.

In the latter case the executable module may be considered as a derived artifact of the source code building process. Furthermore, the building process itself can be described by using the same ontology terms. Indeed, the building process is a kind of execution where a compiler (or a building system like *make*, *ant* or *maven*) is an executable entity while a source code is considered to be the input data for a compiler.

3   Constructing a Software Provisioning Ontology

In order to be capable to describe typical scenarios of CLI software automatic deployment we defined a subject domain ontology *"Automated CLI application*

---

[6] Cartridges encapsulate application components, for example, language runtimes, middleware, and databases, and expose them as services.

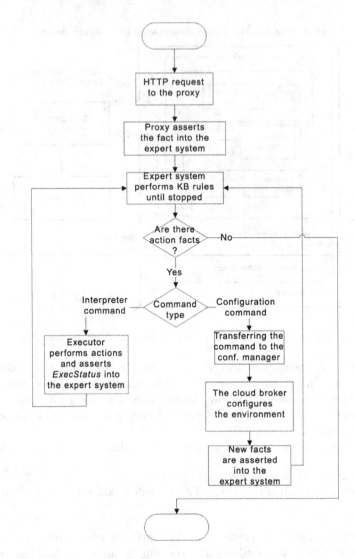

Fig. 3. Main stages of the deployment process

build and run with resolving runtime environment configuration errors" (for further references we use an abridged name *"Software provisioning ontology"*).

3.1   Requirements That Compete: A Case Study

Following Gruber's ontology definition where an ontology is considered as an explicit specification of the conceptualization [21], there are five major ontology design requirements [22]:

1. **Clarity:** An ontology should operate with the intended meanings and with complete (as possible) definitions of introduced terms.
2. **Coherence:** The defined axioms should be logically consistent and all the assertions inferred from the axioms shouldnt contradict to informally given definitions or examples.
3. **Extendibility:** An ontology should be designed so as to allow using shared vocabularies and to support monotonic ontology extension or/and specialization (i.e. new terms might be introduced without revising existing definitions).
4. **Minimal Encoding Bias:** The conceptualization should be specified at the knowledge level without depending on a particular symbol-level encoding.
5. **Minimal Ontological Commitment:** An ontology should require the minimal ontological commitment sufficient to support the intended knowledge sharing activities.

These requirements are competitive: they might contradict to each other. If we consider using ontology term vocabularies in the domain of production knowledge bases, the requirements (1), (3) and (5) are in opposition. If the ontology terms are too common (term commonness being the easiest way to minimize ontological commitment), they are hardly usable to define production rules (since they might produce too much ambiguity). To overcome term ambiguity we have either to revise existing definitions (which contradicts the extendibility principle) or to introduce new terms with complex semantics (which, in turn, contradicts the clarity principle).

Let us consider the following situation: some software module M should be executed triple times (in no particular order) with different arguments $A1$, $A2$, $A3$. Suppose that in order to run the module M with the arguments $A1$ some component K in version V_1 is required (we use the name $K1$ for that). In order to run the module M with the arguments $A2$ the same component K is required, but in version V_2 (the $K2$ component). In order to run the module M with the arguments $A3$ the component K has not to be used at all. The $K1$ and $K2$ configurations are mutually exclusive (i.e the module can not be executed with dependencies on both $K1$ and $K2$). At the beginning (e.g. when the module is being loaded) all the mentioned dependencies are unknown.

Let us introduce facts $NeedK1$ and $NeedK2$ that declare correspondingly that the component configuration $K1$ (for the fact $NeedK1$) or $K2$ (for the fact $NeedK2$) is required in order to successfully execute the module M.

Suppose there are knowledge base rules $R1$ and $R2$ which define how to discover dependencies on $K1$ and $K2$ by analyzing existing installation or execution logs. As a result of executing the production rules $R1$ and $R2$, the facts $NeedK1$ and $NeedK2$ are asserted into the inference engine working memory. Suppose there is also a rule R controlling the module M execution depending on the presence of the facts $NeedK1$ and $NeedK2$ with corresponding dependencies $K1$ and $K2$. Suppose there is a rule R' which is activated when there is no $NeedK1$ nor $NeedK2$ (both facts are absent) in the working memory. Thuswise,

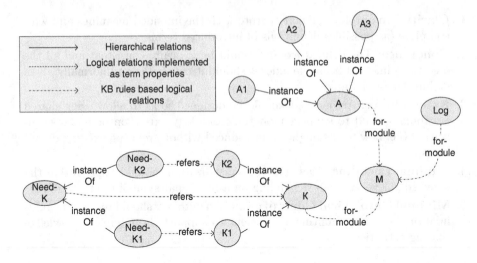

Fig. 4. Running a software module: first attempt to define an ontology

the rule R' controls execution of the module M without dependencies related to the component K.

In fact, such a formal model describes a usual problem of compiling and configuring software projects built from a source code. The different build systems (e.g. *maven* or *ant*) can serve as examples of the module M, target projects can serve as examples of builder arguments A, while third party libraries instantiate dependencies on some component K.

The simple ontology shown in Figure 4 describes the subject domain as follows: there is an A class representing *arguments*. The ontology entities $A1$, $A2$ and $A3$ are instances of A. Class $NeedK$ describes the fact that some component K is required in order to execute the module M. Entities $NeedK1$ and $NeedK2$ are instances of $NeedK$ class containing references to the $K1$ and $K2$ instances of K class. For the reason that the K instances are used with the module M, the K class has an attribute containing a reference to M. The *Log* entity has two attributes: the log *text* and a *reference* to the respective module.

Straightforwardly, Listing 1 demonstrates how to define possible knowledge base rules to be used with this rather naïve ontology.

Listing 1. Knowledge base rules to run a software module (naïve approach)

```
R1:
  if
    Log(text contains ''dependency K1 missed'')
  then
    assert NeedK1()
end
```

```
R2:
  if
    Log(text contains ''dependency K2 missed'')
  then
    assert NeedK2()
end

R:
  if
    $dependency: (NeedK1() or NeedK2())
    $arguments: (A1 or A2 or A3)
  then
    run M with $dependency and $arguments
end

R':
  if
    not NeedK1()
    not NeedK2()
    $arguments: (A1 or A2 or A3)
  then
    run M with $arguments
end
```

It is evident that the ontological model shown in Figure 4 is insufficient to control correct configuration and execution of the module M. There are at least two problems. First, after a sequence of unsuccessful runs (e.g. after an attempt to run the module M with the arguments $A1$ and with no dependencies on K followed by an attempt to run the module M with the arguments $A2$ and with no dependencies on K) the working memory might contain both $NeedK1$ and $NeedK2$ facts. Formally, one can expect that the module shall run with both dependencies ($K1$ and $K2$) which contradicts to the restriction that the components $K1$ and $K2$ are mutually exclusive. Second, the given rules might yield infinite recursion as Table 1 illustrates for one possible series of steps for the subsequently inserted $A1$, $A2$ and $A3$ arguments (in Table 1 the recent items asserted to the working memory being shown bold).

In table 1 rows 7.2 and 7.3 activating the rule R implies executing the module (which has already been successfully run) with beforehand wrong dependencies. Moreover, the step 7 produces recursive rule activation. In the following step 8 the rules $R1$ and $R2$ are inevitably activated again, they assert the facts $NeedK1$ and $NeedK2$, and therefore the rule R is activated four times for every combination of $\{A1, A2\}$ an $\{NeedK1, NeedK2\}$, and then the rules $R1$ and $R2$ are activated all over again.

The latter problem can be fixed by the following improvements of the rules $R1$ and $R2$ (see Listing 2), but this modification doesn't lead to avoiding the unnecessary and irrelevant steps with wrong dependencies.

Table 1. Activating production rules: an issue of infinite recursion

Step	Rule	Facts and Arguments	Working Memory
0	–	–	**A1**
1	R'	A1	A1, **Log("dependency K1 missed")**
2	R1	Log	A1, Log("dependency K1 missed"), **NeedK1**
3	R	NeedK1, A1	A1, Log("dependency K1 missed"), NeedK1, **Log("success")**
4	–	–	A1, Log("dependency K1 missed"), NeedK1, Log("success"), **A2**
5	R	NeedK1, A2	Log("dependency K1 missed"), NeedK1, Log("success"), A2, **Log("dependency K2 missed")**
6	R2	Log	Log("dependency K1 missed"), NeedK1, Log("success"), A2, Log("dependency K2 missed"), **NeedK2**
7.1	R	NeedK2, A2	Log("dependency K1 missed"), NeedK1, Log("success"), A2, Log("dependency K2 missed"), NeedK2, **Log("success")**
7.2	R	NeedK1, A2	Log("dependency K1 missed"), NeedK1, Log("success"), A2, Log("dependency K2 missed"), NeedK2, Log("success"), **Log("dependency K2 missed")**
7.3	R	NeedK2, A1	Log("dependency K1 missed"), NeedK1, Log("success"), A2, Log("dependency K2 missed"), NeedK2, Log("success"), Log("dependency K2 missed"), **Log("dependency K1 missed")**

Listing 2. Fixing the recursion issue

```
R1:
  if
    Log(text contains ''dependency K1 missed'')
    not NeedK1()
  then
    assert NeedK1()
end

R2:
  if
    Log(text contains ''dependency K2 missed'')
    not NeedK2()
  then
    assert NeedK2()
end
```

Furthermore, as Table 2 (steps 9.1 and 9.2) shows, due to the facts $NeedK1$ and $NeedK2$ the rule R' responsible for executing the module M without any external dependency (arguments $A3$) can not be activated. The problem can be solved by adding new rules to the knowledge base in order to allow removing

Table 2. Activating production rules: an issue of irrelevant steps

Step	Rule	Facts and Arguments	Working Memory
8	–	–	Log("dependency K1 missed"), NeedK1, Log("success"), A2, Log("dependency K2 missed"), NeedK2, Log("success"), Log("dependency K2 missed"), Log("dependency K1 missed"), **A3**
9.1	R	NeedK1, A3	Log("dependency K1 missed"), NeedK1, Log("success"), A2, Log("dependency K2 missed"), NeedK2, Log("success"), Log("dependency K2 missed"), Log("dependency K1 missed"), A3, **Log("invalid dependency K1")**
9.2	R	NeedK2, A3	Log("dependency K1 missed"), NeedK1, Log("success"), A2, Log("dependency K2 missed"), NeedK2, Log("success"), Log("dependency K2 missed"), Log("dependency K1 missed"), A3, Log("invalid dependency K1"), **Log("invalid dependency K2")**

the facts $NeedK1$ and $NeedK2$ from the working memory if necessary. Listing 3 fixes this problem.

Listing 3. Rules for running a module with no external dependencies

```
R11:
  if
    Log(text contains ''invalid dependency K1'')
    $dependency: NeedK1()
  then
    remove $dependency
end

R12:
  if
    Log(text contains ''invalid dependency K2'')
    $dependency: NeedK2()
  then
    remove $dependency
end
```

Nonetheless, because of new rules $R11$ and $R12$ appeared, improper rules activation sequences might produce an infinite cycle (consider the sequence $\mathbf{R'(A1)}$, $R1$, $R(A3, NeedK1)$, $R11$, $\mathbf{R'(A1)}$ for instance). Hence, while constructing a rule for inserting facts we have to pay attention to the implementation of other rules. Dependencies between different rules significantly complicate the knowledge base design: rules become unevident and self-contradictory.

The question is how to struggle with this complexity?

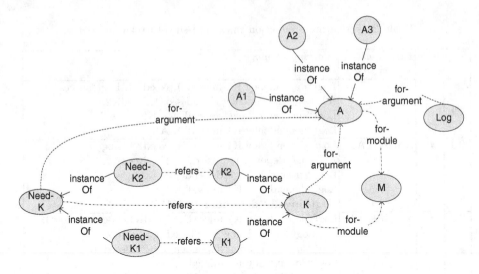

Fig. 5. Running a software module: the revised ontology

Let's turn back to the beginning of the mentioned example and revise the proposed ontology model. We defined the facts $NeedK1$ and $NeedK2$ (and then used them in rules $R1$ and $R2$) as *global scope facts*. It means that they are inserted to the working memory regardless of their connections with the arguments. In reality, the fact say $NeedK1$ has sense only for the concrete arguments. It means that the fact $NeedK1$ should be interpreted not as *"executing the module M with K1 dependency"* but as *"executing the module M with K1 dependency for arguments A1"* or *"executing M with K1 dependency for arguments A2"*, etc. Thus, these facts are **context dependent** and therefore **an ontology should provide concepts supporting the context representation**. It means that we should define an association between the arguments A and the facts $NeedK$ as well as between the Log and the class A. The revised sketch is shown in Figure 5.

The rules corresponding to the just mentioned changes are shown in Listing 4.

The class A represents the context: every predicate within the production rule scope refers to the same parameter $arguments. Semantics of the classes $NeedK$ and Log changed: now there is a relationship linking them to the class A.

Listing 4. Knowledge base rules for the revised ontology

```
R1:
  if
    $arguments: A()
    Log($arguments, text contains ''dependency K1 missed'')
  then
    assert NeedK1($arguments)
end
```

```
R2:
  if
    $arguments: A()
    Log($arguments, text contains ''dependency K2 missed'')
  then
    assert NeedK2($arguments)
end

R:
  if
    $arguments: A()
    $dependency: (NeedK1($arguments) or NeedK2($arguments))
  then
    run M with $dependency and $arguments
end

R':
  if
    $arguments: A()
    not NeedK1($arguments)
    not NeedK2($arguments)
  then
    run M with $arguments
end
```

For the sake of compactness we skip here further analysis of the sample ontology design which allows us to discover the connection between the *Log* class and the component *K* as well as the rule priority issues.

Nevertheless, even this simplified example implies an important consideration. Paying attention to the competing principles mentioned in the beginning of this section we have to admit that designing subject domain ontology (which is targeted to be used in order to construct ontologies of specific tasks) is a complex problem requiring both solid working experience and much intuition. Indeed, for software engineering purposes one of the most important criteria whether an ontology is successful or not depends on the question whether it is consistent and flexible enough to be used to solve practical problems of software configuration and execution (i.e. only minor extensions of a base ontology are required in order to construct the derived ontologies of specific tasks).

3.2 Core Concepts: Activities, Requests and Related Facts

The software provisioning ontology major concepts are *activities* and activity *requests*. These concepts might aggregate each other: a request might consist of activities, and vice versa, an activity might aggregate requests (being subrequests, in a sense). An *Activity* is a sequence of *actions* provided by an expert or by a user in order to achieve an activity goal, while a *Request* might be considered as a new goal setting.

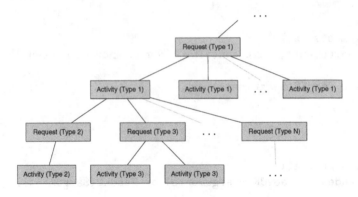

Fig. 6. A hierarchy of activities and requests

Formally, *Activity/Request* hierarchies can be represented by using trees (see Figure 6).

In order to describe activity results, we introduce a concept of an *Activity status* which is twofold: there is an *Activity runtime status* and there is an *Activity completion status*. The *Activity runtime status* instances are an *Activity being executed* and an *Activity suspended*. The *Activity completion status* instances are an *Activity succeeded* and an *Activity failed*. Assume that an activity is completed successfully if the activity goal is reached (for example, for the activity *Unpacking* the artifact has been successfully unpacked), otherwise the activity is failed (for example, some file artifact has not been unpacked for the reason that the required archiving utility has not been found).

The *Request* features a necessary and appropriate condition for starting the activity. For each request instance the activity caused by this request should be known.

Similar to an activity concept, a *Request* might also have its status which is also twofold: there is a *Request runtime status* and there is a *Request completion status*. The *Request runtime status* instances are a *Request being executed* and a *Request suspended* while the *Request completion status* instances are a *Request succeeded* and a *Request failed*. If at least one activity for the request is completed successfully, the request is considered to be completed successfully too. By contrast, if all the activities associated with the given request failed, the request is considered to be failed.

Since there might be many objects in the inference engine working memory and these objects might activate many rules at once, assuring rules consistency is a difficult problem. One of the possible ways to solve this problem is to design an ontology in a way that big groups of rules are never activated at once, while all the activated rules are easily computable. In order to support such an approach we introduce the following concepts (see also Figure 7):

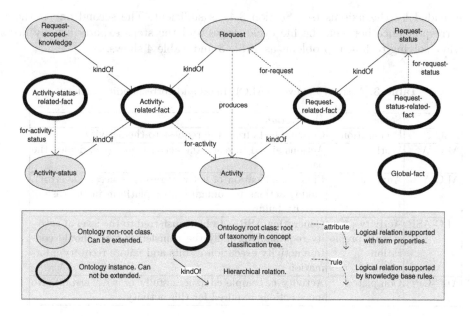

Fig. 7. Activities and requests are major ontology concepts

- An **Activity related fact (ARF)** is a base concept for every fact corresponding to a certain activity. This class has a reference $ARF.activity$ to an activity instance. Within the context of executing some activity Ac only those facts should be considered which are either global scope facts or the facts related directly to the activity Ac (i.e. $ARF.activity = Ac$).
- A **Request related fact (RRF)** is a base concept for every fact corresponding to a certain request. This class has a reference $RRF.request$ to a request instance. Within the context of executing some request Rq only those facts should be considered which are either global scope facts or the facts related directly to the request Rq (i.e. $RRF.request = Rq$).
- An **Activity status related fact(ASRF)** is similar to an ARF but related to the activity status.
- A **Request status related fact (RSRF)** is similar to an RRF but related to the request status.
- An **Activity life cycle state (ALCS)** is a base class required to describe the activity life cycle states.
- A **Request life cycle state (RLCS)** is a base class required to describe the request life cycle states.

Let us explain the reasons why the activity and request life cycle stages concepts are required. The first reason is production rules simplification. In most cases in order to define rules for the specific tasks (e.g. building with *maven*, running *Java* application, etc.) only rules for stages ALC-W and ALC-A (introduced in Table 3) have to be defined. For other stages the default behavior provided by the axioms (see Section 3.4) is sufficient. The second one is the correspondence between the life cycle stages and the steps experts usually do while solving technical problems as Table 3 and Table 4 shows.

Table 3. Activity life cycle (ALC) stages and corresponding actions

Code	Stage	Description
ALC-P	Preparation	Copying facts from the request to the activity
ALC-W	Work	Actions aimed to reach the activity goal (e.g. deploy the package).
ALC-A	Analysis	Final classification of activity results (success or error). Taking actions to configure host platform in the case of activity failure.
ALC-ASP	Activity status preparation	Generating ASRFs to be transferred to the level of activity requests. Normally the transferred facts should contain activity execution results and failure recovery information
ALC-C	Completed	Activity is completed (successfully or with errors). No more actions required for this activity

Table 4. Request life cycle (RLC) stages and corresponding actions

Code	Stage	Description
RLC-P	Preparation	Copying facts from the parent activity to its subrequest
RLC-M	Main	Performing actions aimed to achieve the request goals: performing an activity and activity status analysis. In case of recoverable errors (which are fixed during ALC-A phase) – re-executing the activity
RLC-RSP	Request status preparation	Generating RSRFs to be transferred to the level of the parent activity. Normally the transferred facts should contain successful request results
RLC-C	Completed	Request is completed. The goal is achieved (probably as a result of several attempts to perform an activity) or not (the request failed). No more actions required for this request

3.3   Aliases, References, and Request Scoped Knowledge

Different activity instances might be (and often should be) *isolated* from each other: analysis of an activity instance doesn't depend on whether other activity instances (of the same or of different types) exist. An activity instance scope is supported by *Activity related facts*: only those facts which are related to this instance should be regarded. Such facts *grouping* allows the *Activity* to represent a fact interpretation context. Nonetheless, there are cases when neither grouping is acceptable, nor isolation. Quite the reverse, it might be required that the facts

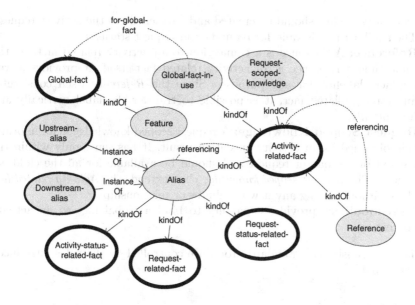

Fig. 8. Aliases, references, and request scoped knowledge representation

are allowed to be transferred from one activity to another. We introduce two ways of such a transfer: *vertical* and *horizontal* transfer (according to the *Activity/Request* graph layout shown in Figure 6). There are four types of vertical facts transfer:

- Transferring facts from the parent *request* to the child *activity* (**RA-down-transfer**): transferring input arguments from the request to the activity.
- Transferring facts from the child *activity* to the parent *request* (**AR-up-transfer**): transferring activity execution results to the parent request.
- Transferring facts from the parent *activity* to the child *request* (**AR-down-transfer**): transferring input arguments from the activity to the request.
- Transferring facts from the child *request* to the parent *activity* (**RA-up-transfer**): transferring (sub)request results to the parent activity.

The horizontal transfer is possible only between the activities within the request scope and implemented by combination of *AR-up-transfer* and *RA-down-transfer*.

To support fact transfers the following concepts are introduced (see also Figure 8):

- **Alias:** A (fact) alias is an object containing a reference to an activity related fact which, in turn, is either a request related fact or an activity related fact or a request status related fact or an activity status related fact. Hence, if an activity needs the fact to be up-transferred, this activity should create an *Upstream alias* for this fact and "attach" the *alias* to an activity status. If an activity needs the fact to be down-transferred (to the subrequest), a

Downstream alias should be created and "attached" to the activity request. The similar thing is valid for requests and request statuses.

- **Reference:** A reference is a connection to an activity related fact. At the same time a reference is an activity related fact itself. It provides a way for facts labeling for any purposes. Since the *Reference* itself constitutes an activity related fact, it is possible to transfer the label vertically and horizontally.
- **Request Scoped Knowledge:** A request scoped knowledge is an information obtained as an activity execution result. It has sense only within the context of the parent request and should be available for all the child activities. The *Request scoped knowledge* class extends the *Activity related fact* class without adding any new attributes or relationships.
- **UseGlobalFact** provides capability to bind a global fact to an activity context

The fact transfer is a standard procedure controlled by the rules introduced in Section 3.4.

3.4   Ontology Axioms

For the introduced concept model we defined the following axiomatic rules:

1. **Activity Life Cycle Axioms**
   (a) For any *activity* there is always only one *fact* describing the *activity life cycle* from the set $\{ALC\text{-}P, ALC\text{-}W, ALC\text{-}A, ALC\text{-}ASP, ALC\text{-}C\}$.
   (b) For any *activity* its life cycle stages follow the only allowed sequence: $ALC\text{-}P$, $ALC\text{-}W$, $ALC\text{-}A$, $ALC\text{-}ASP$, $ALC\text{-}C$. The transitions are possible only in the moments when there is no active knowledge base rules, and no other operation on action facts is performed and no incomplete subrequests (i.e. subrequests with a status different from $RLC\text{-}C$).

2. **Request Life Cycle Axioms**
   (a) For any *request* there is always only one *fact* describing the *request life cycle* from the set $\{RLC\text{-}P, RLC\text{-}M, RLC\text{-}RSP, RLC\text{-}C\}$.
   (b) For any *request* its life cycle stages follow the only allowed sequence: $RLC\text{-}P$, $RLC\text{-}M$, $RLC\text{-}RSP$, $RLC\text{-}C$. The transitions are possible only in the moments when there is no active knowledge base rules, and no other operation on action facts is performed and no incomplete child activities (i.e. child activities with a status different from $ALC\text{-}C$).

3. **Vertical Fact Transfer Axioms**
   (a) **RA-Down-Transfer Axiom.** For any *request* being in the stage $RLC\text{-}M$ and any *child activity* being in the stage $ALC\text{-}P$ the following is required: for any *Downstream alias* "attached" to the *request* the object that this *alias* refers to should be copied and "attached" to the *activity*.
   (b) **AR-Up-Transfer Axiom.** For any *request* being in the stage $RLC\text{-}M$ and any *child activity* being in the stage $ALC\text{-}ASP$ the following is required: for any *Upstream alias* "attached" to the *activity status* this *alias* should be copied and "attached" to the *request status*.

(c) **AR-Down-Transfer Axiom.** For any *activity* being in the stage *ALC-W* or *ALC-A* and any *child request* being in the stage *RLC-P* the following is possible: for any *fact* "attached" to the *activity* a *Downstream alias* can be created and "attached" to the *request*.

(d) **RA-Up-Transfer Axiom.** For any *activity* being in the stage *ALC-W* or *ALC-A* and any *child request* being in the stage *RLC-RSP* the following is required: for any *Upstream alias* "attached" to the *request status* the object that this *alias* refers to should be copied and "attached" to the *activity*.

4. **Horizontal Fact Transfer Axiom:** For any *request* being in the stage *RLC-M* and any *child activity* being in the stage *ALC-ASP* the following is required: for any *request scoped knowledge* fact created in the activity stages ALC-W or ALC-A and "attached" to the *activity* the *Downstream alias* should be created and attached to the *request*. Note that in contrast to the *AR-up-transfer* axiom, the *alias* is "attached" to the *request* itself, not to the *request status*. With using the *RA-down-transfer* axiom, it means that the *request scoped knowledge* is transferred to all new *activities*.

5. **Axiom about Creating an Activity in Response to the Request:** For any *request* if there is no child activity at all or if all the existing *child activities* are in the stage *ALC-C* and all the *child activities* are completed with error, and for every *child activity* at least one *ErrorFixed* fact exists, then a new *activity* should be created and "attached" to the current *request*.

6. **Request Status Axioms**

(a) **Successful Request Axiom:** For any *request* if all the *child activities* are in the stage *ALC-C* and there is at least one *child activity* with status *activity succeeded*, then the *request* is completed with status *request succeeded*.

(b) **Failed Request Axiom:** For any *request* if all the *child activities* are in the stage *ALC-C*, there is no any *child activity* with status *activity succeeded*, and conditions of the axiom (5) are not met, the *request* is completed with status *request error*.

The ontology axioms define rules of transferring facts between requests and activities. The axioms guarantee that an activity is re-performed (within the context of the same request) until either the activity goal is reached or all known error resolving procedures are examined.

3.5 Actions, Common Facts and Global Facts

The action facts are used to arrange indirect communications between a knowledge base and an execution environment. The environment executes the required actions and asserts the action status facts back into the working memory.

The important point is that action fact definitions allow retaining the *declarative* (non-imperative) form of the knowledge base rules. For example, if it is required to execute an application, the knowledge base simply asserts an *ExecAction* fact to be performed by the executor (instead of running the application by itself). However, asserting facts into the inference engine working memory does affect neither an environment nor an expert system except the case that new rules might be activated in the knowledge base. In the earlier described target architecture (see Figure 2) there is a special component aimed to process action facts – the executor. The implementation of how the executor interacts with the expert system can affect the implementation of the knowledge base rules. For our implementation we consider a model of deferred action execution (where action facts are executed if and only if there is no more any active rule left in the inference engine schedule). In the moment when there is no more any rule scheduled for the inference engine, the executor takes control. The executor checks whether there are the action facts in the working memory. As soon as the action is executed, the new facts representing execution results are asserted into the working memory. The processed facts are then removed from the working memory.

We introduced several common actions as Figure 9 shows. Major action facts in the software provisioning ontology are the following:

1. **ExecAction:** This fact represents a shell command to be executed. The execution result is asserted into the working memory in the form of an *ExecStatus* fact.
2. **AddFeatureAction:** This fact means that the execution environment needs to be changed (some external components should be added or deleted). The *AddFeatureAction* is an abstract entity. Its subclasses should be defined (for example, a subclass *AddJDK7*) in order to communicate properly with the configuration manager.
3. **UserAction:** This fact represents a requirement that some action has to be performed by a user

Common facts (see also Figure 9) are interpreted within the context of some activity or some request (i.e. common facts are related fact instances). In contrast, global facts are context independent. Major common facts are represented by the *Artifact*, *ExecCommand* and *UserInfo* classes. Here is the explanations of major common facts:

1. **Artifact** is an artifact in the local file system, e.g. a file (*FileArtifact*) or a folder (*FolderArtifact*).
2. **ArtifactRef** is an artifact (specified by a URL) in the remote file system.
3. **ExecCommand** describes a command line interpreter command. The command format includes a program name and positional and key/value arguments. The command is executed under control of the deployment manager agent in response to the action fact *ExecAction*.
4. **UserInfo** is an arbitrary information to be shown to a user.
5. **FileArtifactList** represents a list of *FileArtifact* facts.

We assume that within the ontologies of specific tasks new common fact types may be added similar to the example as we demonstrate in section 4.

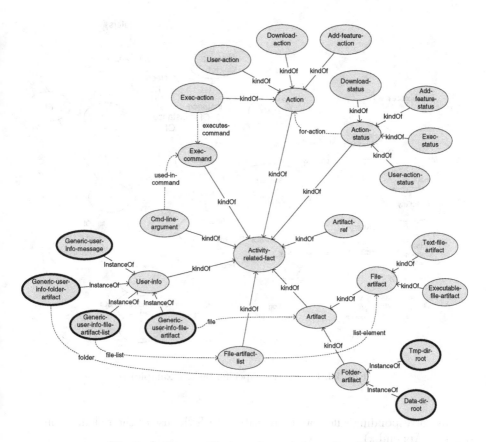

Fig. 9. Actions, artifacts and user information facts

4   Designing an Ontology of Specific Tasks

Subject domain ontologies are rarely used in expert systems directly. The reason is that such formalisms are usually too common to describe the subject domain related specific tasks. However, we are able to define an ontology of specific tasks by extending base entities of the core ontology.

4.1   Running a Module Example Revisited

With respect to the introduced concepts of the core ontology, the ontology of specific tasks for the above mentioned example of running some software module M can be revised as Figure 10 shows.

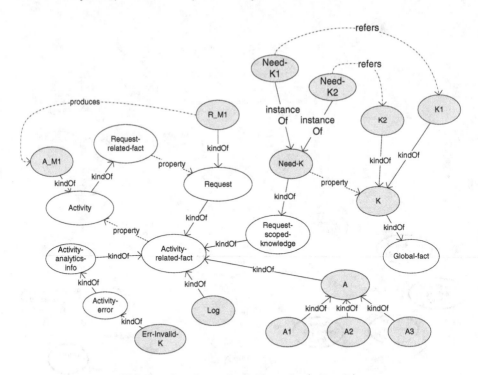

Fig. 10. Ontology of specific tasks (example)

The corresponding rules can be rewritten with the use of the redefined ontology (see Appendix).

4.2   Example of *maven* Build and *Java* Programs Execution

Figure 11 and Figure 12 demonstrate how to use the software provisioning ontology to describe the task of building an application with using *maven* build system and the task of executing a Java application

As you can see, the core ontology concepts are descriptive enough to serve as a foundation for definition of relatively complex derived specific ontologies without introducing many new terms and without revision of existing concepts. Let us note that for the sake of paper readability we skip the further detailed demonstration on how to construct the knowledge base production rules in order to manage processes of client application building and execution with detecting respective errors while using some building tool (e.g. *maven*) as a kind of specific building system.

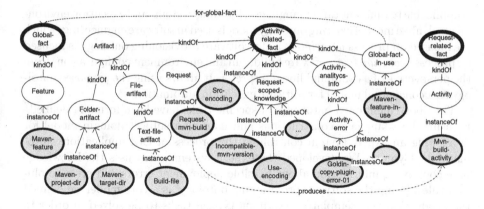

Fig. 11. An ontology of specific tasks: building applications with using *maven*

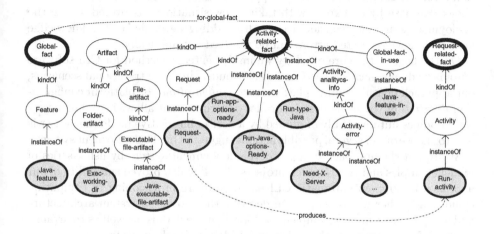

Fig. 12. An ontology of specific tasks: executing *Java* applications

5   Conclusion

In this work we defined an ontology which provides a conceptual core for building a networked environment for provisioning software applications to computing clouds and for resolving execution environment configuration errors automatically. We introduced an architecture for application deployment automation in computing clouds. By using a series of examples related to the software engineering practices and with respect to the requirements of research software we examined a question why designing an ontology is a complex problem and analyzed major iterations in the process of ontology construction.

Since the proposed general-purpose ontology can't be used directly to define expert system knowledge base rules, we demonstrated how the ontologies of specific tasks can be designed within the subject domain of building software projects with *maven* and their execution in the Java runtime environment. The general schema is applicable to different target languages, building systems and execution environments. Unlike most of existing ontology models used in software engineering (which are focused on process *description*), the software provisioning ontology is focused on process *execution* (e.g. software build, run and environment configuration) from the perspective of a command line interpreter. It allows using the proposed ontology as a conceptual model for software execution automation.

The above mentioned ontologies of specific tasks have been formally defined[7] by using Java language constructs and used as a subject domain model for developing an expert system controlling the process of CLI applications automatic deployment. We developed the expert system which uses the knowledge base containing information about possible deployment errors and error resolution rules. Using the software provisioning ontology as a core model we defined the production rule templates describing typical tasks to be solved in order to execute the required software and to identify possible execution and configuration errors. It is important that the approach allows further modifications of the knowledge base by an expert with taking new situations discovered during the deployment stage into consideration. After adding new rules to the knowledge base such newly detected errors can be resolved automatically.

On the base of the core ontology we implemented a method for CLI software automatic deployment in *PaaS* and *IaaS* clouds. Unlike to traditional scenarios, our approach doesn't require platform *pre-configuration* nor platform configuration *description* (by using descriptor files or scripts), but allows installing necessary modules automatically or guided by a user in interactive mode (i.e. a user is able to choose one of several possible actions suggested by the deployment system).

We tested the approach in a prototype deployment system by using series of model examples and two research projects [23,24] which proved suitability of our approach to support automatic CLI-based software provisioning to computing clouds. In further works we plan to describe the prototype system architecture and a deployment manager implementation in more details, as well as to arrange a series of experiments with a selection of MIR research projects.

References

1. Mirex home, `http://www.music-ir.org/mirex/wiki/mirex_home`
2. Cannam, C., Benetos, E., Mauch, M., Davies, M.E., Dixon, S., Landone, C., Noland, K., Stowell, D.: Mirex 2014: Vamp plugins from the centre for digital music
3. West, K., Kumar, A., Shirk, A., Zhu, G., Downie, J., Ehmann, A., Bay, M.: The networked environment for music analysis (nema). In: 2010 6th World Congress on Services (SERVICES-1), pp. 314–317 (July 2010)
4. Bunch, C.: Automated Configuration and Deployment of Applications in Heterogeneous Cloud Environments. PhD thesis, Santa Barbara, CA, USA, AAI3553710 (2012)
5. Kuznetsov, A., Pyshkin, E.: An ontology of software building, execution and environment configuration and its application for software deployment in computing clouds. St. Petersburg State Polytechnical University Journal. Computer Science. Telecommunications and Control Systems 2(193), 110–125 (2014)

---

[7] `https://github.com/andrei-kuznetsov/fpf4mir/tree/master/fpf4mir-core`

6. Rabkin, A., Katz, R.: Precomputing possible configuration error diagnoses. In: 2011 26th IEEE/ACM International Conference on Automated Software Engineering (ASE), pp. 193–202 (November 2011)
7. Zhang, S.: Confdiagnoser: An automated configuration error diagnosis tool for java software. In: Proceedings of the 2013 International Conference on Software Engineering, ICSE 2013, pp. 1438–1440. IEEE Press, Piscataway (2013)
8. Dong, Z., Ghanavati, M., Andrzejak, A.: Automated diagnosis of software misconfigurations based on static analysis. In: 2013 IEEE International Symposium on Software Reliability Engineering Workshops (ISSREW), pp. 162–168 (November 2013)
9. Ya-Yunn, S., Attariyan, M., Flinn, J.: Autobash: Improving configuration management with operating system causality analysis. In: Proceedings of the 21st ACM Symposium on Operating Systems Principles, pp. 237–250. Stevenson (2007)
10. Attariyan, M., Flinn, J.: Automating configuration troubleshooting with dynamic information flow analysis. In: Proceedings of the 9th USENIX Conference on Operating Systems Design and Implementation, OSDI 2010, pp. 1–11. USENIX Association, Berkeley (2010)
11. Zhao, Y., Dong, J., Peng, T.: Ontology classification for semantic-web-based software engineering. IEEE Trans. Serv. Comput. 2(4), 303–317 (2009)
12. Liao, L., Qu, Y., Leung, H.: A software process ontology and its application. In: First Intl. Workshop on Semantic Web Enabled Software Eng. (November 2005)
13. Caralt, J., Kim, J.W.: Ontology driven requirements query. In: 40th Annual Hawaii International Conference on System Sciences, HICSS 2007, pp. 197c–197c (January 2007)
14. Ambrosio, A., de Santos, D., de Lucena, F., da Silva, J.: Software engineering documentation: an ontology-based approach. In: Proceedings of the WebMedia and LA-Web, pp. 38–40 (October 2004)
15. Shahri, H.H., Hendler, J.A., Porter, A.A.: Software configuration management using ontologies (2007)
16. Psl core, http://www.mel.nist.gov/psl/psl-ontology/psl_core.html
17. Web services business process execution language version 2.0 (oasis standard April 11, 2007), http://docs.oasis-open.org/wsbpel/2.0/os/wsbpel-v2.0-os.html
18. Business process model and notation version 2.0 (bpmn 2.0) (omg standard January 02, 2011), http://www.omg.org/spec/bpmn/2.0/pdf/
19. Aitken, S.: Process representation and planning in cyc: From scripts and scenes to constraints (2001)
20. Pyshkin, E., Kuznetsov, A.: A provisioning service for automatic command line applications deployment in computing clouds. In: IEEE Proceedings of the 16th IEEE Conference on High-Performance Computing and Communications, pp. 526–529 (2014)
21. Gruber, T.R.: A translation approach to portable ontology specifications. Knowl. Acquis. 5(2), 199–220 (1993)
22. Gruber, T.R.: Toward principles for the design of ontologies used for knowledge sharing. Int. J. Hum.-Comput. Stud. 43(5-6), 907–928 (1995)
23. Glazyrin, N.: Audio chord estimation using chroma reduced spectrogram and self-similarity. In: Proceedings of the Music Information Retrieval Evaluation Exchange, MIREX (2012)
24. Khadkevich, M., Omologo, M.: Large-scale cover song identification using chord profiles. In: de Souza Britto Jr., A., Gouyon, F., Dixon, S. (eds.) ISMIR, pp. 233–238 (2013)

Appendix: Knowledge Base Rules Revised According to the Edited Ontology of Specific Tasks

```
EI:
  if  // Error identification
    $activity : A_M1()
    ALCWork(activity == $activity)
    Log(activity == $activity, text contains ''dependency K[0-9] missed'')
  then
    assert ErrInvalidK($activity)
    assert ActivityFailed($activity)
    logger.print(''Activity failed because of missed component K'')
end

SI:
  if  // Success identification
    $activity : A_M1()
    ALCWork(activity == $activity)
    Log(activity == $activity, text contains ''success'')
  then
    assert ActivitySucceeded($activity)
    logger.print(''Activity succeeded'')
end

R1:
  if  // need K1
    $activity : A_M1()
    ALCAnalyze(activity == $activity)
    $err : ErrInvalidK(activity == $activity)
    Log(activity == $activity, text contains ''dependency K1 missed'')
  then
    assert NeedK1($activity)
    assert ActivityErrorFixed($activity, $err)
    logger.print(''Problem fixed - using K1 next time'')
end

R2:
  if  // need K2
    $activity : A_M1()
    ALCAnalyze(activity == $activity)
    $err : ErrInvalidK(activity == $activity)
    Log(activity == $activity, text contains ''dependency K2 missed'')
  then
    assert NeedK2($activity)
    assert ActivityErrorFixed($activity, $err)
```

```
      logger.print(''Problem fixed - using K2 next time'')
end

R':
  if  // run with no dependencies
    $activity : A_M1()
    ALCWork(activity == $activity)
    $a : A(activity == $activity)
    not NeedK(activity == $activity)
  then
    //run M with $arguments
    $cmd := ExecCommand($activity, M, $arguments)
    assert ExecAction($activity, $cmd)
end

R:
  if  // run with dependencies
    $activity : A_M1()
    ALCWork(activity == $activity)
    $a : A(activity == $activity)
    $dependency: NeedK(activity == $activity)
  then
    //run M with $dependency and $arguments
    $cmd := ExecCommand($activity, M, $dependency, $arguments)
    assert ExecAction($activity, $cmd)
end
```

*AIDA: A Language
of Big Information Resources

Yutaka Watanobe and Nikolay Mirenkov

University of Aizu, Aizu-wakamatsu, Fukushima 965-8580, Japan
{yutaka,nikmir}@u-aizu.ac.jp

Abstract. Some features of *AIDA language and its environment are
provided to show a way for possible preparing well-organized information
resources which are based on integrated-data architecture supporting
searching, understanding and immediate re-use of the resources needed.
A project of big information resources of the above mentioned type is
presented and relations of users and resource unit owners within Global
Knowledge Market are briefly considered. Some ideas behind knowledge
and experience transfer with permanent re-evaluating resource unit values and examples of the resource types are also provided.

1 Introduction

The potential of big data concept is really great. There are a number of successful
stories about big data of exhaust- web-search type applied in communication,
leisure, and commerce (see, for example [1-2]). The concept is especially promoted by companies when they want something to sell. In many cases, data-rich
models of a theory-free approach are based on discovering statistical correlations
(statistical patterns in the data) rather than causations. Such discovering is not
so expensive and can be (depending on application) accurate enough. However,
the idea that "the numbers reliably speak for themselves" meets a lot of criticism
because without understanding what is a basis for a correlation it is difficult to
realize a possible reason for the correlation disappearance [3-5]. Some backlash
about the effectiveness of big data is expressed in [3]: "Big Data has arrived, but
big insights have not." In many cases, big data sets are messy, non-transparent,
and difficult for finding what sampling biases hide inside them. In addition, big
data systems can be easily gamed. On the other hand, bad analysis of big data
and inappropriate actions do not mean that there are no examples where good
analysis combined with large volumes of data has been applied to gain good insight. In any case, Big Data is a new tool and we should learn a lot to understand
how to efficiently use it [4-5].

Big information resources are also big data; the difference is in semantic rich-
ness of data components and their compositions. Though statistical patterns
can also be discovered within information resources, a primary goal of manipulation with resource units is in understanding of how to make decisions based on
knowledge and experience of others and how to present your decision to others.

W. Chu et al. (Eds.): DNIS 2015, LNCS 8999, pp. 112–121, 2015.

In other words, it is to acquire and transfer knowledge and experience between people (including people of different generations [6]). We promote the concept of big information resources as large-scale sets of well-organized, reliable, and accurate units representing people knowledge and experience applicable for education, research, and business.

In this paper we present some features of *AIDA language and its environment for preparing well-organized information resources which are based on integrated-data architecture of "self-explanatory features"; those are features to be convenient for people searching, understanding and immediate use of the resources needed. We also present a project of big information resources which units with corresponding ownerships are acquired within the framework of Global Knowledge Market where knowledge and experience transfer is permanently performed and resource unit values are re-evaluated.

2   Related Work and Motivation

Our work presented here has relations with many trends in research and development, aimed to exploit hybrid intelligence of software components, visual languages applied for developing such components and interfaces to access them. We are interested in methods and techniques for analyzing semantic data to discover connections, in intelligence summarization methods, exploratory search, visualization and navigation, users' modelling, annotation, adaptation, and feedback (especially, within research and educational activities) [7-12].

In addition, we are interested in methods and related systems for acquiring big sets of such components. In particular, ideas employed for organizing wiki-type systems (and, first of all, Wikipedia [13]) should be mentioned. This includes key human factors that influence the exploration of large interconnected complex data and user-supportive environments helping users in discovering connections and understanding meaning of different aspects of large volumes of data.

Though deploying big data into environment is important for us [1-6], here we focus on cognitive and intelligent aspects of the component development and on motivation of people involved in creating reliable and accurate units of information resources [6,13-14]. This includes how people share their insight and experience, make decisions based on such sharing and how integration of information within a resource unit is performed and big information resources are practically appeared. We are also interested in why MOOCs initiative, in spite on top level experts involved, is not so successful [15].

People cognition abilities, in a great part, depend on their abilities to perform mental simulation. However, this simulation depends not only on internal information processing mechanisms of the brain, but also on levels of abstraction connecting the physical world things and corresponding abstract models. To make such simulation more efficient, it is necessary to decrease spending of people energy and time for the recognition of semantics behind unknown terminology, for understanding associations between objects which syntax-semantics forms are based on rote memorization and for fighting with concepts and notations that have a trend to be over-abstracted. In other words, disorientation and cognitive overload are serious obstacles for people. They become tired after overcoming such obstacles and do not have enough energy and time for focusing on their own ideas and models, and on the use of their real knowledge and cognitive abilities [16].

3   Programming in Pictures

Programming in pictures (we also call it as Filmification of Methods) is a special approach to help people for some saving their energy and time. It is an approach where pictures and moving pictures are used as super-characters for representing features of computational algorithms and data structures, as well as for explaining models and application methods involved. *AIDA is a language supporting programming in pictures. Within this approach some "data space structures" are traversed by "fronts of computation" and/or some "units of activity" are traversed by flows of data. There are compound pictures to define algorithmic steps (called Algorithmic CyberFrames) and generic pictures to define the contents of compound pictures. Compound pictures are assembled into special series to represent some algorithmic features. The series are assembled into Algorithmic Cyberscenes and CyberFilm. The generic/compound pictures and their series are developed and acquired in special galleries of an open type where supportive pictures of embedded clarity annotations are also included.

*AIDA programs are presented as a set of information resources where application people can present not only application requirements, specifications and program texts, but also various features of models applied, ideas behind, methods involved, etc. In other words, *AIDA programs are information resources which can be used for different goals and based on an idea of comprehensive explanations and automatic generations of corresponding documents, presentation slides, executable codes, etc.

In fact, such information resources can be a basis for a global environment of active knowledge which can be much easier to search, understand, and immediately reuse. They can be applied for new schemes of decision making processes based on knowledge and experience of others acquired in the global environment. They can also be a basis of educational materials of a new generation. There are many publications where various explanations and comparisons of *AIDA features and experiments related are presented. A selected list of these publications can be found at [17]. Though *AIDA and its environment are an experimental system, it includes a necessary set of components and subsystems to support application people in developing and representing models, application algorithms, and related information resources for various types of involvements. Fig.1 illustrates some system features.

An extended set of applications where some experiments based on *AIDA have been done is depicted by Fig. 2 super-symbols. Though, one of the goals of *AIDA initiative has been creating a new language of the technical literature, now we have realized that it can also be used for preparing other types of information resources.

Activities of application people

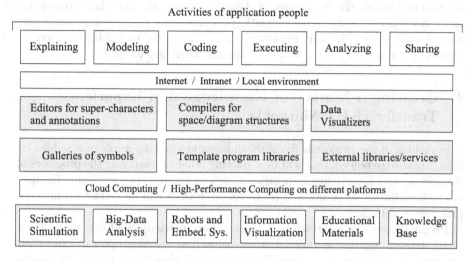

Fig. 1. *AIDA components and subsystems for supporting various applications

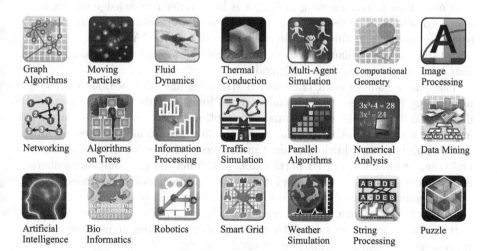

Fig. 2. Possible application fields

To create and maintain the global environment of reliable and accurate information resources, the free style of Wikipedia is not appropriate. On the one hand, a serious refereeing process should be involved, and on the other, the motivation of people to create the resources should be high. That is why we propose a project of big information resources which units with corresponding ownerships are acquired within the framework of global knowledge market where Knowledge and Experience Transfer is permanently performed and resource unit values are re-evaluated. This is to start exploration and mining the global information resources and marketing them in a style similar to oil, gas or gold.

4   Creating a Market of Knowledge and Experience Transfer (KET Market)

A basic idea of the project is to establish a center (playing a role of a publisher, data/knowledge house), that will be a starting pillar of managing the processes of Knowledge and Experience acquisition, assessment, and transfer. Some features of stock exchange, lotteries, and pachinko operations can be embedded into the processes mentioned.

People, possessing interesting and important knowledge and experience (first of all, experts in some application areas, as well as senior people) prepare their knowledge and experience (including sweat and wisdom) in a form of special information resources for transferring to other people (first of all, to beginners or younger generation, but also to other experts). To start, two types of units of information resources are introduced. The first type is related to the algorithm representation in *AIDA language. In this case, an algorithm is considered as a plan of actions to reach a goal; in fact, it is how transfer knowledge and experience into actions. The second type is related to the representation of a vision of some life realities. In this case, the vision can also be in *AIDA language, but with the focus on explanation through multiple views and annotations, rather than on generating any executable code.

The center collects such information resources and after serious refereeing and editing (in a sense, as for a conventional journal) puts them into a special distributed database. A conventional subscription for access to the resources is arranged and some ratings of the resource units are obtained based on subscriber opinions. In addition, a few times a week, the center holds assessment and transfer sessions in order participants of the market (not only subscribers) could demonstrate their real-time interests to the information resources.

The demonstration of the interest to a resource (the importance of a resource for people) can be done by participants in a form of rating evaluation through putting some amount of money for certain information resources or in a form requesting a partial or exclusive copy right transfer.

The market mechanism can be based on the following. Information resources involved in session (the involvement is based on transparent rules) obtain a rating depending on money put on them. Among these resources, shares are selected and participants who demonstrated their interest to these resources obtain some money return which can be much higher than their investment used for the rating evaluation. In fact, the amount of return is calculated by some standard rules where all money collected are distributed between the center, to cover spending and get some profit, some possible supportive organizations, successful participants and resource creators to award their activities. Rating evaluation of the resources is combined from the current rating of subscribers and the rating of the latest assessment session. It can be used as an orientation for the market participants during the next assessment sessions. In addition to the rating, some part of money collected can be preserved for the next sessions.

A fundamental point of any information resources market is how to attractively explain features of corresponding resources and, on the other hand, to protect owner rights. Multiple view format of *AIDA programs (Fig.3) is a very good basis to solve this issue. It allows easily to introduce a few levels of access applied for all users (a promotion level), for subscribers (a level differentiating user's interests), and for market players (defining exclusive rights). An example of such levels can be recognized in deep green, green, and light green colors of supporting grounds.

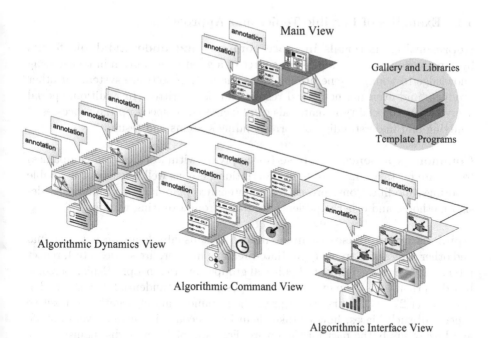

Fig. 3. *AIDA program format

5 Types of Information Resources for Transferring Knowledge and Experience

As we have mentioned in the previous section, the first type of information resource is related to the algorithm representation in *AIDA language (including application models, methods behind them, supporting software, clarifying annotations, etc.) and the second type is related to the representation of a vision of some life realities. Hereafter we focus on some details of the second type resources which forms can be: questionnaires filled out, classifications developed, observations and evidences of the best practice provided, opinions expressed, advices recommended, overviews/tutorials implemented, essays prepared, memoirs and books written, albums and gallery/library of pictures, sounds, movies and/or other items collected, databases based on systematic sets of facts and approaches for domain-specific problem solving, methods (tools) developed to get experience acquired from experts (adults, relatives, etc.). An exciting example of memoirs writing activity is provided by Italian City of Diaries (Pieve Santo Stefano) where more than 7,000 diaries, letters, autobiographies, and punctilious notes present the lives of common people [6].

5.1 Examples of Possible Topics and Approaches

Representing materials in more compact and understandable forms by experience persons is a possible source of a great contribution in transferring knowledge to younger generations. Many manuals of software systems or other products and even mathematical results can be rewritten. In addition, special overviews or tutorial type materials can be created to speed up the processes of studying and understanding of corresponding systems and results.

Common experience database (for applying within or between companies) to acquire mistakes, failures, fault descriptions, etc. as well as successful trouble shootings, modifications, pro-active maintenance planning, benchmarking, design methods, and other experience feedback from a retiring workforce.

Opinions about biases of mass-media that usually focus on hot, criminal and other topics of extreme types (late-breaking murder, fire stories, etc.), reflect intentions of owners, influential political groups, and rich people. Senior persons, based on their long-term experience and possibly on confidential documents disclosed after 25 or 30 years, can express their opinions about negative or positive aspects of such biases, how to take them into account in current everyday life and how to apply the past for the future. For example, how to distinguish really your opinion (based on a variety of facts and sources of information) and opinion you can consider as yours because of mass-media influence.

Personal evidences and feelings about the atomic bombing and world war: the number of people who have got such evidences and direct involvement is rapidly decreased. So, it is an urgent issue to collect their memories and feelings in order to transfer them to future generations and to avoid the lost and deformation of facts and/or their interpretation.
In a similar way, many things related to Dai-ichi nuclear plant and 2011-tsunami can be prepared.

Grandparent or senior teacher recommendations to young parents about 1) how to influence children to control impulses, focus attention and regulate emotions (schools spend a lot of time to work with students who cannot do these things), 2) how to organize the relations that children have outside school (these relations shape children performance inside the school), 3) how to establish attentive, attuned parental relationships in family (family relations are matter more than anything else: early childhood attachments shape lifelong learning competence, help to choose friends wisely, handle frustration better, and be more resilient in face of setbacks).

Nurses and mature women recommendations to young mothers about how to establish healthier attachments. They should give to young mothers the sort of cajoling and practical wisdom that in other times would have been delivered by grandmothers.

Right questions in right time are an important thing to create useful information resources. Now many questionnaires or polls are for political, administrative or mass-media goals. They do not take into account real users' needs, problems, wishes, feelings, plans, etc. For example, how many people in the country have headache every morning and how many people have feeling of unhappiness going to bed? Such information can be extremely important for government or industry organizers but it is not available now. The senior people can point to many such important questions about evaluation of current life aspects and right time of asking them.

No mistake first versus taking a risk is an important alternative to be understood. **No mistake first** is a Japanese style of activity, while **taking a risk** is more popular in the USA and other western countries. Providing knowledge and specific facts for understanding this alternative and all possible aspects of taking a risk can be a task for the seniors.

A Great Variety of Other Topics and Approaches Can Be Mentioned

- Conflicts with children: what is behind?
- Bullying: now and in the past
- Preserving and developing traditions
- Reading to acquire experience
- Gaming to acquire experience
- Developing projects to acquire experience
- Writing essays on people relations
- Writing essays on businesspersons motivation
- Writing essays on politicians spirit
- Personal observations: my travels abroad
- My connections and partners
- Creativity and bureaucracy
- Collectivism and individualism

6   Summary and Conclusions

The center of the KET market will become a global "publisher" and data warehouse acquiring knowledge and experience of people from all over the world. Any person can make contribution into this "intelligence of humanity" and become an owner of some units of information resources. Searching and reasoning on big information resources (well-organized, reliable, and accurate units) can automatically take into account a great variety of the resource features and their relations including space data structures, types of data, units of their measurement, algorithmic scenes, high-level operations, model descriptions, results of experiments, multiple views, etc. as well multilevel clarifying (semantic-based) annotations including their keywords. These features will help in clustering people, objects and processes based on patterns of similarity, difference, efficiency, dual and emotional aspects, frequency, and scalability, positions in space/time, personality, and satisfaction.

Discovering patterns of such types is a basis for the second level of the big information resources analytics; this is for introducing various classification schemes in which points of different directions of the classification space are represented by different patterns, compound patterns, or different values of the patterns. Clustering (in fact, meta-clustering) in this classification space, including discovering empty spots in it, is promoted as a way to recommend or at least to hint about new possible features and innovation decisions. They allow specifying new forms of analytics workflow for discovering various relations and correlations between software components, periods of their developments and revisions, people and methods involved, successes and failures in applications, personal reactions and behaviors, etc.

Systematic re-evaluation of the resource unit importance by market players will positively influence on quality of application decisions. This will drastically change the values of information resources and big data analytics, and a new information ecosystem will be established. As a result, these will speedup processes of discovering new phenomena, creating new technologies, and introducing more efficient educational materials and learning methods.

References

1. Big Data and Analytics Success Stories:
   http://www.csc.com/big_data/success_stories
2. Big Data Success: 3 Companies Share Secrets:
   http://www.informationweek.com/big-data/big-data-analytics/
   big-data-success-3-companies-share-secrets/d/d-id/1111815?
3. Harford, T.: Big data: are we making a big mistake? FT Magazine (March 28, 2014), http://www.ft.com/intl/cms/s/2/21a6e7d8-b479-11e3-a09a00144feabdc0.html#axzz34wlErtQx
4. Dodson, S.: Big Data, Big Hype? Innovation Insights (April 25, 2014), http://innovationinsights.wired.com/insights/2014/04/big-data-big-hype/
5. Marcus, G., Davis, E.: Eight (No, Nine!) Problems With Big Data, New York Times (April 6, 2014)
6. Povoledo, E.: A Trove of Diaries Meant to Be Read by Others (August 19, 2014), http://www.nytimes.com/2014/08/20/world/europe/a-trove-of-diaries-meant-to-be-read-by-others.html?_r=0
7. Shaffer, C.A., Cooper, M.L., Alon, A.J.D., Akbar, M., Stewart, M., Ponce, S., Edwards, S.H.: Algorithm visualization: the state of the field. ACM Transactions on Computing Education, 10, 1-22 (August 2010)
8. AlgoViz.org: The algorithm visualization portal, http://algoviz.org/ (accessed: March 26, 2014)
9. Dann, W.P., Cooper, S., Pausch, R.: Learning To Program with Alice, 2nd ed. Prentice Hall Press, Upper addle River (2008)
10. Ford, J.L.: Scratch Programming for Teens, 1st ed. Course Technology Press, Boston (2008)
11. Introduction to Lego Robotics with RCX Code Programming Language, http://robofest.net/academy/rcxcode4robofest.pdf
12. Bitter, R., Mohiuddin, T., Nawrocki, M.: LabView: Advanced Programming Techniques, 2nd edn. CRC Press (2007)
13. Wikipedia, the free encyclopedia, http://en.wikipedia.org/wiki/Big_data
14. Watanobe, Y., Mirenkov, N., Terasaka, H.: Information resources of *AIDA programs. In: The Proceedings of the IEEE Symposium on Visual Languages and Human Centric Computing, Melbourne (July-August 2014)
15. Pope, J.: What Are MOOCs Good For? (December 15, 2014), http://www.technologyreview.com/review/533406/what-are-moocs-good-for/
16. Watanobe, Y., Mirenkov, N.: Hybrid intelligence aspects of programming in *AIDA algorithmic pictures. Future Generation Computer Systems 37, 417–428 (2014)
17. *AIDA language: http://aida.u-aizu.ac.jp/aida/index.jsp

Interactive Tweaking
of Text Analytics Dashboards

Arnab Nandi, Ziqi Huang, Man Cao, Micha Elsner, Lilong Jiang,
Srinivasan Parthasarathy, and Ramiya Venkatachalam*

The Ohio State University
Columbus, Ohio, USA

Abstract. With the increasing importance of text analytics in all disciplines, e.g., science, business, and social media analytics, it has become important to extract actionable insights from text in a timely manner. Insights from text analytics are conventionally presented as visualizations and dashboards to the analyst. While these insights are intended to be set up as a one-time task and observed in a passive manner, most use cases in the real world require constant tweaking of these dashboards in order to adapt to new data analysis settings. Current systems supporting such analysis have grown from simplistic chains of aggregations to complex pipelines with a range of implicit (or latent) and explicit parametric knobs. The re-execution of such pipelines can be computationally expensive, and the increased query-response time at each step may significantly delay the analysis task. Enabling the analyst to interactively tweak and explore the space allows the analyst to get a better hold on the data and insights. We propose a novel interactive framework that allows social media analysts to tweak the text mining dashboards not just during its development stage, but also *during* the analytics process itself. Our framework leverages opportunities unique to text pipelines to ensure fast response times, allowing for a smooth, rich and usable exploration of an entire analytics space.

Keywords: text analytics, interactivity, database systems, social media analysis.

1 Introduction

Both the sciences and commercial enterprises are increasingly turning to analyzing textual data such as social media to enable them to act on intelligence gleaned from the data. This, in turn, has led to an increasingly complex set of analysis tools and frameworks [2, 12, 17, 33]. Given the scale of data in these contexts (e.g., for social media, over 500 million tweets are sent each day [35] and over 1.35 billion – almost half [15] of the adult Internet population – is on

---

* Now at IBM Silicon Valley Labs. Work done while at Ohio State.
This work was supported in part by NSF under grant IIS-1422977.

W. Chu et al. (Eds.): DNIS 2015, LNCS 8999, pp. 122–132, 2015.

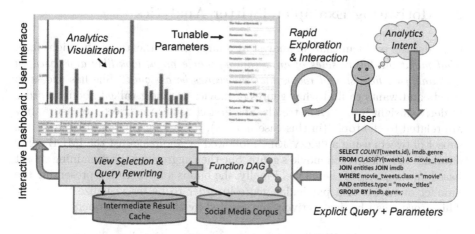

Fig. 1. The IntPipe System for exploratory and interactive tweaking of text pipelines: User specifies an analytics dashboard as an explicit query, which is executed over a social media corpus and then presented as a visualization dashboard in the user interface. The user then explores and interacts with parameters triggering further queries, which are rewritten to leverage the intermediate result cache.

Facebook [13]), it is critical to rethink the text mining process in the context of exploratory analytics for efficiently surfacing time critical insights on text data.

An analytics task in this space is typically composed of a set of ordered and incremental text processing components to generate insights. A typical text analytics pipeline includes multiple stages of text preprocessing, model-building and evaluation, and aggregation. These pipelines can get fairly complex to manage as they often involve a stack of third party tools within a massive framework. Additionally, each step in the pipeline requires the tuning of several explicit and/or implicit (latent) parameters. The notion of "latent" is due to the fact that these parameters are tunable but not always exposed to the end user for exploration. For example, the normalization process of text has tunable parameters like change case/remove punctuations. However, these components are currently a black box to the end user and the user does not often have any grasp on these tunable knobs through the interface. Rigid text mining pipelines, with a plethora of these black box components, often hinder the analytics process from being interactive.

In this work we describe IntPipe, a framework that enables exploratory analytics through the interactive tweaking of text pipelines. The goals of our framework are orthogonal to existing analytics suites – we focus on the interactive aspects of tweaking text pipelines. Our framework exposes the different parameters (both latent and explicit) employed at each stage of the text pipeline to allow users explore and tweak.

2   Motivating Example: Twitter Analytics

Consider the problem that an analyst at a large movie theater chain might face: *what might the future sales of an upcoming movie might look like; and based on that analysis, how should rooms and showtimes be assigned?* She has access to tweets and wants to find what genres of movies are most talked about in order to derive insight into current trends. She first classifies the tweets to ensure they are related to the topic (in this case *movies*), extracts named entities from this subset of tweets using a FUZZY JOIN[1], followed by a join against the IMDB genre database to look up each movie's genre. She then aggregates the remaining subset based on the genre of the movie. Finally, she builds a histogram representing the number of tweets of a particular genre in order to better visualize the result. This process can be declaratively represented by the SQL-style query:

Query 1.1. Distribution of movie genres mentioned on Twitter

```
SELECT COUNT(movie_tweets.id), imdb.genre
FROM CLASSIFY(tweets) AS movie_tweets
JOIN entities
ON movie_tweets.id = entities.id
JOIN imdb ON entities.title = imdb.title
GROUP BY imdb.genre;
```

Going forward, the analyst may notice that there is an unusual distribution of tweets in a particular genre. She can then dive into this genre to manually inspect a list of tweets that have been classified under this genre and reclassify those that have been wrongly classified as movies. This reclassification will not change the overall query above, but will involve a change in the training data used for prediction and the resulting model is now used to reclassify the data. This *tweaked* version of the prior query can be represented as follows:

Query 1.2. Retraining and reclassifying existing tweets

```
SELECT COUNT(movie_tweets.id), imdb.genre
FROM CLASSIFY(tweets, new_training_data) AS movie_tweets
JOIN entities
ON movie_tweets.id = entities.id
JOIN imdb ON entities.title = imdb.title
GROUP BY imdb.genre;
```

While performing the query above, the analyst may also notice that one reason for the unusual behavior is that in some cases tweets that have numbers (which are also names of movie titles) are wrongly included as movies. For example, a tweet whose text is "Let's meet at 9" will be considered as a tweet containing the movie title "9" since "9" is a movie title. This can be fixed by changing a parameter in the title extraction of the stemming and preprocessing stage to exclude numbers.

---

[1] We use the *fuzzy* or approximate string join [9, 18] as a standard operation, orthogonal to the contributions of this paper.

This query can be represented as follows, the *is_number* controls whether we need to handle the movie whose name is a number:

Query 1.3. Applying all previous queries and then handle movie titles as numbers

```
SELECT COUNT(movie_tweets.id), imdb.genre
FROM CLASSIFY(tweets, new_training_data) AS movie_tweets
JOIN entities
ON movie_tweets.id = entities.id
AND NOT entities.is_number()
JOIN imdb ON entities.title = imdb.title
GROUP BY imdb.genre;
```

By analyzing the sample tweets of a specific genre, she may discover that the reason for the incorrect movie title is that some movie titles were reformatted as hashtags. Hashtags, which users prefer to use to highlight words, are very popular in Twitter. Therefore, the analyst can unwrap the movie title in order to get the correct movie title. The query can further be modified as follows:

Query 1.4. Applying all previous functions and then unwrapping hashtags to movie titles

```
SELECT COUNT(movie_tweets.id), imdb.genre
FROM CLASSIFY(tweets, new_training_data) AS movie_tweets
JOIN entities
ON movie_tweets.id = entities.id
AND NOT entities.is_number()
AND entities.hashtag().is_title()
JOIN imdb ON entities.title = imdb.title
GROUP BY imdb.genre;
```

By looking at the sample tweets for a specific genre, she may realize that some movie titles did not get extracted. This commonly happens in social media, due to the casual writing style. In order match the movie title better, the analyst needs to normalize both the tweets and movie titles in the IMDB (or any other movie database).

Query 1.5. Applying all previous functions and then change tweets to lower case

```
SELECT COUNT(movie_tweets.id), imdb.genre
FROM CLASSIFY(tweets, new_training_data) AS movie_tweets
JOIN lowercase(entities) as entities
ON movie_tweets.id = entities.id
AND NOT entities.is_number()
AND entities.hashtag().is_title()
JOIN imdb ON entities.title = imdb.title
GROUP BY imdb.genre;
```

Now with these popular movie titles mentioned in tweets, the analyst may want to look into temporal trends, filtering movies by their release dates:

Query 1.6. Applying all previous functions and then filter movies by its release date

```
SELECT COUNT(movie_tweets.id), imdb.genre
FROM CLASSIFY(tweets, new_training_data) AS movie_tweets
JOIN lowercase(entities) as entities
ON movie_tweets.id = entities.id
AND NOT entities.is_number()
AND entities.hashtag().is_title()
JOIN imdb ON entities.title = imdb.title
AND year(imdb.date) = "2013"
GROUP BY imdb.genre;
```

Observing all of the queries listed above, we can see that each query is only a small modification of the original query, resulting in the final query tree shown below. Therefore, there are opportunities of reusing the results of the prior query. For example, Query 1.2 just needs to rebuild the classifier model and relabel all tweets, then filter the tweets that have been labeled as 1 – there is no need to re-do the normalization, title, and feature extraction steps.

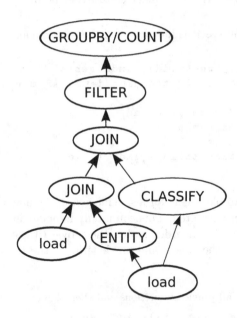

Fig. 2. Query tree representing the final example query

Given the motivation and running example, we now envision an interactive dashboard for text analytics. The following section describes the challenges faced when building such a framework, and Section 4 details IntPipe, our system.

3   Challenges

Motivated by this example, the vision of IntPipe is to provides an interactive dashboard for text analytics which can **respond to the end-user within interactive times**. To this end, we list two key issues faced with this vision.

3.1   Interactive Response Times

Empirical evaluations in human-computer interaction have demonstrated that increased response times influence data exploration behavior [20], and that response times in the 100-1000ms window [24, 34] allow interfaces to seem perceptibly "instantaneous" with respect to end-user interaction. Pipelines, and the following analytical steps are strictly batch operations, typically performed in a one-time setting. After the initial pipeline creation phase, there is little support to iteration and interaction with the text pipeline, or tweaking various parts of the text processing, performing subtle but significant changes. For a given query, each combination of parameters represents a unique processing of data. Given that the interface exposed to users is expected to be *interactive*, each change in query or tuning parameter should affect a change in the analytics output in low latency. In the simple case, however, a change in the query, or any one of the parameters triggers a re-computation of the entire analytics pipeline. Given the possible resource requirements, complexity of each of the steps involved, and the parameter space, analyzing data at scale by recomputing the entire pipeline is thus untenable.

3.2   Text Operations as Blackboxes

Tools for natural language processing that form the core of the pipeline are typically black boxes. The underlying mining process is not transparent to end users. With large scale, ad-hoc text analytics coming into play, this text mining process can become slow and arduous. Thus, exploration of various parameter settings will be time-consuming as the entire pipeline has to be re-executed every time for every follow-up query involving a new parameter setting. Also, the opaque nature of such functions makes it impossible to perform query optimizations such as predicate pushdowns. It is thus necessary to expose the parameters of these black boxes and make them configurable by providing a dynamic dashboard for interactive exploration.

As a solution to these challenges, we propose to expedite this process, and allow analysts to *interactively* modify large-scale text pipelines. Enabling such interactivity has several compelling uses. End users can tweak and overhaul text pipelines not just during the inception stage of the pipeline, but *during* the analytics process itself. Further, it allows for interactive ad-hoc analysis of text and the ability to modify the pipeline to correctly articulate the overall analytical intent. Finally, it allows for quick "what-if" analyses of the data, testing several hypothetical assumptions on the dataset.

4   The IntPipe Approach

Based on the motivating example and the challenges at hand, we propose three methods that enable the interactive tweaking of text pipelines, collectively named IntPipe. As shown in Figure 1, IntPipe models queries as DAGs (directed acyclic graphs) of text processing functions, and leverages an intermediate result cache to speed up text analytics queries.

4.1   Trading Off Computation and Cache

A key observation of the analytics process is that follow-up queries are typically a slight modification of the prior query. Thus, based on this interaction pattern, there will likely be several opportunities for result reuse. Reusing intermediate results allows us to drastically reduce the execution needs of each analytics pipeline, thereby making interactive response times possible. There is however, an important caveat: caching *all* the intermediate results can be prohibitively expensive (e.g., one version of the entire corpus per function node or more, if the function inflates the data, such as shingling).

We are thus left with an interesting problem: given a particular query pipeline, for which of the subqueries should we cache intermediate results, and which should we discard? Our insight into this problem is to *consider all the possible transformations of the current query, and cache subqueries which provide the most savings of execution time for this transformed query, given the least additional space requirement*. We model this problem as an optimization problem for a query D as follows:

$$\underset{subqueries \in \text{POWER-SET}(D.subqueries)}{\arg\max} \quad \sum_{D' \in Transforms(D)} Benefit(D') \quad (1)$$

where:

$$Benefit(D') = t \cdot Speedup(D', subqueries) - d \cdot Size(D', subqueries) \quad (2)$$

Here $D.subqueries$ represents the set of subqueries in the query, and TRANS-FORMS(D) represents the set of queries that can be constructed by modifying the current query D, either by changing a parameter or the structure of the query. *Speedup* and *Size* represent the time benefit estimate and the data usage estimate for caching the additional intermediate results at a *subquery*. Estimators and parameters can be empirically derived by profiling the functions over samples of the data.

4.2   Reusability

Unlike traditional result reuse, text pipelines have an interesting property that needs elucidation. The output of certain text transformation functions can be **reusable** with respect to another function, i.e., we can materialize the result of queries with old parameter values and reuse it to evaluate the query with new parameter values. For example, consider `DISTINCT(TO_UPPER(entities))`. Should the analyst decide that the `TO_UPPER` be changed to `TO_LOWER` for some reason, instead of triggering a full re-execution, it is possible to *reuse* the prior result, and simply return `TO_LOWER(DISTINCT(TO_UPPER(entities)))`, which is logically equivalent since `TO_LOWER` and `TO_UPPER` are reusable for each other. This optimization is particularly useful for large collections with a small number of distinct elements. Upon recognizing such an opportunity, IntPipe can trigger a query rewrite based on the result cache contents, drastically reducing response time. It should be noted that not all functions are reusable – thus the pairwise compatibility of functions will need to be considered at the query rewrite layer.

4.3 Opening Black Boxes

A key challenge mentioned in Section 3 is that some text operations are typically viewed as *black boxes*, not amenable to optimizations such as cache reuse. To show that we can indeed make text operations amenable to optimizations, we demonstrate simple modifications to *one* such black box – a Naive Bayes classifier – that allow for caching and incremental use. We pick Naive Bayes as an example since it works well in several domains, its performance and robustness are well understood [5, 23, 25, 30], and since incremental variants have been widely studied [7, 19, 26]. Other classifiers and faster methods are heavily motivated [4, 6, 8, 14, 25] and can be similarly modified to reuse the intermediate result cache.

The probabilistic model for a Naive Bayes classifier is a conditional probabilistic model:

$$P(C|F_1, ...F_n) = \frac{P(C) \prod_i P(F_i|C)}{P(F_1, ...F_n)} \tag{3}$$

More formally, given $P(C)$, the probability of two classes: `movie tweet` / `non-movie tweet`, and $P(F_i|C)$, the conditional probability of one feature value given the class, we can compute the probability of $P(C|F_1, ...F_n)$, the conditional probability of one class given the feature vector.

Initial Naive Bayes Classifier Model: We now consider caching opportunities within the classifier, when implementing the classifier inside a relational database. The basis of a Naive Bayes model involves maintaining a histogram of feature occurrences in the training set, a relatively fast database aggregation operation. The speed also allows for easy updates, in case the user supplies additional new training data. Further, a database implementation allows for the persistence and reusability of intermediate and final tables created for classification.

Implementing Incremental Computation in Naive Bayes: Since a Naive Bayes Classifier is a simple probabilistic classifier based on applying Bayes theorem with strong (naive) independence assumptions, for each new training data with feature vector $F = (F_1 = 1, F_2 = 0, ...F_n = 0)$ and $C = C_1$, we merely need to update the count of $P(F_1 = 1, C = C_1)$, $P(F_2 = 0, C = C_1)$ and all feature values in the feature vector. However, a small update of the count will not change the final result. Therefore, we set a threshold to filter out the small updates of count and only change the count in the likelihood table when the update is larger than the threshold. Since our use case is that of social media where tweets have a 140-character limitation, the feature vector of each tweet will be very sparse. We expect the number of changes in the likelihood table to be small, and thus the number of database (and cache) updates will be small. This can be implemented by setting up a chain of database triggers that batch count updates from the likelihood table to a hash value table to the prediction table. Changes are propagated once the updates cross pre-fixed thresholds.

5   Related Work

Visualization-based analytics dashboards for text and social media have gathered increased attention [1, 2] lately. The IntPipe framework explores making such systems more interactive and tunable, by introducing parameterized exploration of the underlying text mining stack itself. Frameworks on social media data have investigated a variety of text-oriented functions beyond typical database processing, such as location analysis [33], semantic analysis [17], topic analysis [29], and prediction [3]. Olston et al. [27], have looked into the iterative analysis of web-scale data using query templates, utilizing a combination of offline and online computation given a query workload. Our framework utilizes a similar formalism and enables the interactive exploration of such analytics by speeding up execution. Marcus et al. [22] introduce a stream-oriented query processing system for Twitter events. Ideas from our work can significantly improve the iterative refinement user flow of such a system.

The concept of *recycling* of intermediate results has also been employed in several different aspects of databases unrelated to text analytics, such as column stores [16]. From the caching perspective, there exists a significant body of work in query rewriting using materialized views [11, 21, 28]. Gupta and Mumik [10] provide the theoretical bounds for selecting views to materialize under a maintenance cost constraint, which we can use as a basis for our optimization: [31] investigates a similar determination of additional views to be materialized as an optimization problem over the space of possible view sets. Roy et al. [32] demonstrates the practicality of heuristic-based, multi-query optimization. These studies above are based on standard SQL functions; our framework additionally considers user-defined functions, and leverages reuse properties specific to text mining pipelines.

6   Conclusion and Future Work

In this paper, we introduced IntPipe – a framework motivated by the growing use of text analytics dashboards. IntPipe allows parameterized tweaking of text processing pipelines, by modeling them as DAGs of possibly reusable functions.

By leveraging intermediate result reuse and query rewriting, IntPipe enables fast and iterative processing of text pipelines.

Going forward, the vision for IntPipe can be extended in three fronts, First, it would be useful to handle not just changing text pipelines, but also continually changing data, in the form of streaming queries. The provision of result reuse in this regard is challenging, given that intermediate results will need to be invalidated as the stream of data passes. This can be made possible by efficient use of database triggers. Second, we observe that the analyst is often faced with too many parameters and knobs to tweak. In this light, in addition to providing interactive access to tweaking pipelines, the system could also *guide* the user to possibly ideal parameter settings. This can be done by speculating, precomputing, and caching parameter combinations ahead of time. Third, following the pattern demonstrated with the Naive Bayes classifier, the framework could extend to support more complex text processing functions such as named entity recognition, clustering and advanced classification methods, most of which are still considered blackboxes from a data processing infrastructure perspective.

References

1. Aggarwal, C.C.: An Introduction to Social Network Data Analytics. Springer (2011)
2. Alexe, B., Hernandez, M.A., Hildrum, K.W., Krishnamurthy, R., Koutrika, G., Nagarajan, M., Roitman, H., Shmueli-Scheuer, M., Stanoi, I.R., Venkatramani, C., Wagle, R.: Surfacing Time-critical Insights from Social Media. In: SIGMOD (2012)
3. Asur, S., Huberman, B.A.: Predicting the Future with Social Media. In: WI-IAT (2010)
4. Deng, K., Moore, A.W.: Multiresolution Instance-based Learning. In: IJCAI (1995)
5. Domingos, P., Pazzani, M.: On the Optimality of the Simple Bayesian Classifier under Zero-One Loss. In: Machine Learning (1997)
6. Fisher, D.H.: Knowledge Acquisition via Incremental Conceptual Clustering. In: Machine Learning (1987)
7. Gama, J.: A Cost-sensitive Iterative bayes. In: ICML (2000)
8. Gama, J., Castillo, G.: Adaptive Bayes. In: Advances in AI BERAMIA (2002)
9. Gravano, L., Ipeirotis, P.G., Jagadish, H.V., Koudas, N., Muthukrishnan, S., Pietarinen, L., Srivastava, D.: Using q-grams in a DBMS for Approximate String Processing. In: TCDE (2001)
10. Gupta, H., Mumick, I.S.: Selection of Views to Materialize in a Data Warehouse. In: TKDE (2005)
11. Halevy, A.Y.: Answering Queries Using Views: A Survey. In: VLDB (2001)
12. Infosphere Biginsights, I. (2011), http://www.ibm.com
13. Facebook Inc. 1.35 Billion Monthly Active Users as of. Company Information (September 30, 2014)
14. Indyk, P., Motwani, R.: Approximate Nearest Neighbors: Towards Removing the Curse of Dimensionality. In: STOC (1998)
15. International Telecommunication Union: United Nations Special Agency. The World in 2014. ICT Facts and Figures (2014)
16. Ivanova, M.G., Kersten, M.L., Nes, N.J.: An Architecture for Recycling Intermediates in a Column-store. In: TODS (2010)

17. Jadhav, A.S., Purohit, H., Kapanipathi, P., Anantharam, P., Ranabahu, A.H., Nguyen, V., Mendes, P.N., Smith, A.G., Cooney, M., Sheth, A.: Twitris 2.0: Semantically Empowered System for Understanding Perceptions from Social Data. In: ISWC (2010)

18. Koudas, N., Marathe, A., Srivastava, D.: Flexible String Matching Against Large Databases in Practice. In: VLDB (2004)

19. Lewis, D.D.: Naive (Bayes) at Forty: The Independence Assumption in Information Retrieval. Springer, 1998

20. Liu, Z., Heer, J.: The effects of interactive latency on exploratory visual analysis. IEEE Trans. Visualization & Comp. Graphics, Proc. InfoVis (2014)

21. Mami, I., Bellahsene, Z.: A Survey of View Selection Methods. In: SIGMOD (2012)

22. Marcus, A., Bernstein, M.S., Badar, O., Karger, D.R., Madden, S., Miller, R.C.: Tweets as Data: Demonstration of TweeQL and Twitinfo. In: SIGMOD (2011)

23. McCallum, A., Nigam, K.: A Comparison of Event Models for naive bayes Text Classification. AAAI-LTC (1998)

24. Miller, R.B.: Response time in man-computer conversational transactions. In: Proceedings of the, Fall Joint Computer Conference, Part I, December 9-11, pp. 267–277. ACM (1968)

25. Moore, A., Lee, M.S.: Cached Sufficient Statistics for Efficient Machine Learning with Large Datasets. JAIR (1998)

26. Murphy, K.P.: Naive Bayes Classifiers. Springer (2006)

27. Olston, C., Bortnikov, E., Elmeleegy, K., Junqueira, F., Reed, B.: Interactive Analysis of Web-scale Data. In: CIDR (2009)

28. Park, C.-S., Kim, M.H., Lee, Y.-J.: Finding an Efficient Rewriting of OLAP Queries Using Materialized Views in Data Warehouses. In: DSS (2002)

29. Reips, U., Garaizar, P.: Mining Twitter: A Source for Psychological Wisdom of the Crowds. Behavior Research Methods (2011)

30. Rish, I.: An Empirical Study of the Naive bayes Classifier. IJCAI (2001)

31. Ross, K.A., Srivastava, D., Sudarshan., S.: Materialized View Maintenance and Integrity Constraint checking: Trading Space for Time. In: SIGMOD (1996)

32. Roy, P., Seshadri, S., Sudarshan, S., Bhobe, S.: Efficient and Extensible Algorithms for Multi Query Optimization. In: SIGMOD (2000)

33. Sankaranarayanan, J., Samet, H., Teitler, B.E., Lieberman, M.D., Sperling, J.: Twitterstand: News in Tweets. SIGSPATIAL GIS (2009)

34. Shneiderman, B.: Response time and display rate in human performance with computers. ACM Computing Surveys (CSUR) 16(3), 265–285 (1984)

35. Twitter Inc. Twitter Usage: 500 million Tweets are sent per day. Company Information (2014)

Topic-Specific YouTube Crawling
to Detect Online Radicalization

Swati Agarwal[1] and Ashish Sureka[2]

[1] Indraprastha Institute of Information Technology-Delhi (IIIT-D), India
swatia@iiitd.ac.in
[2] Software Analytics Research Lab (SARL), India
ashish@iiitd.ac.in

Abstract. Online video sharing platforms such as YouTube contains several videos and users promoting hate and extremism. Due to low barrier to publication and anonymity, YouTube is misused as a platform by some users and communities to post negative videos disseminating hatred against a particular religion, country or person. We formulate the problem of identification of such malicious videos as a search problem and present a focused-crawler based approach consisting of various components performing several tasks: search strategy or algorithm, node similarity computation metric, learning from exemplary profiles serving as training data, stopping criterion, node classifier and queue manager. We implement two versions of the focused crawler: best-first search and shark search. We conduct a series of experiments by varying the seed, number of n-grams in the language model based comparer, similarity threshold for the classifier and present the results of the experiments using standard Information Retrieval metrics such as precision, recall and F-measure. The accuracy of the proposed solution on the sample dataset is 69% and 74% for the best-first and shark search respectively. We perform characterization study (by manual and visual inspection) of the anti-India hate and extremism promoting videos retrieved by the focused crawler based on terms present in the title of the videos, YouTube category, average length of videos, content focus and target audience. We present the result of applying Social Network Analysis based measures to extract communities and identify core and influential users.

Keywords: Mining User Generated Content, Social Media Analytics, Information Retrieval, Focused Crawler, Social Network Analysis, Hate and Extremism Detection, Video Sharing Website, Online Radicalization.

1 Research Motivation and Aim

YouTube is a most popular video sharing website that allows users to watch and upload an unlimited number of videos. It also allows users to interact with each other by performing many social networking activities. According to YouTube statistics[1], over 6 billion hours of video are watched each month on YouTube.

---

[1] http://www.youtube.com/yt/press/statistics.html

W. Chu et al. (Eds.): DNIS 2015, LNCS 8999, pp. 133–151, 2015.

100 millions of people perform social activities every week and millions of new subscriptions are made every day. These subscriptions allow a user to connect to other users[2]. The high reachability of videos among users (videos are easily accessible to viewers for free, without the need of an account), low publication barriers (users need only a valid YouTube account) and anonymity (their identity is unknown) has led users to misuse YouTube in many ways by uploading malignant content that are offensive and illegal. For example, harassment and insulting videos [19], video spam [16], pornographic content [3], hate promoting [17] and copyright infringed videos [2].

Research shows that YouTube has become a convenient platform for many hate and extremist groups to share information and promote their ideologies. The reason because video is the most usable medium to share views with others [6]. Previous studies show that extremist groups put forth hateful speech, offensive comments and messages focusing their mission [11]. Social networking allows these users (uploading extremist videos, posting violent comments, subscribers of these channels) to facilitate recruitment, gradually reaching world wide viewers, connecting to other hate promoting groups, spreading extremist content and forming their communities sharing a common agenda [7] [20].

Online radicalization and extremism have a major impact on society that contributes to the crime against humanity[3]. The presence of such extremist content in large amount is a major concern for YouTube moderators (to uphold the reputation of the website), government and law enforcement agencies (identifying extremist content and user communities to stop such promotion in country). However, despite several community guidelines and administrative efforts made by YouTube, it has become a repository of large amounts of malicious and offensive videos [17]. Detecting such hate promoting videos and users is significant and technically challenging problem. 100 hours of videos are uploaded every minute, that makes YouTube a very dynamic website. Hence, locating such users by keyword based search is overwhelmingly impractical. The work presented in this paper is motivated by the need of a solution to combat and counter online radicalization. We frame our problem as 1) identifying such videos and users, promoting hate and extremism (Focus of this paper) on YouTube, 2) locating virtual and hidden communities of hate promoting users sharing a common goal or group mission and 3) identifying users with strong connections and playing central role in a community.

The research aim of the work presented in this paper is the following

1. To investigate the application of a focused crawler (best first search and shark search) based approach for retrieving YouTube user-profiles promoting hate and extremism. Our aim is to examine the effectiveness of two versions of the focused crawler (best-first search and shark search) and measure performance by varying experimental parameters such as the size of the n-gram, similarity threshold and seed.

---

[2] http://www.jeffbullas.com/2012/05/23/

[3] http://curiosity.discovery.com/question/how-hate-crime-impact-society

Table 1. Summary of Literature Survey of 14 Papers, Arranged in Reverse Chronological Order, Identifying Hate & Extremist Content on Various Platforms. VS= Video Sharing Websites, MB= Micro-Blogging, BL- Blogging, SN= Social Networking, DF= Discussion Forum, OW= Other Websites.

S.No.	Research Study	Platform	Objective & Analysis
1.	O'Callaghan et. al; 2013	MB	Analysis of extreme right activities on multiple platforms for community detection.
2.	I-Hsien Ting et. al; 2013	SN	Identifying extremist groups on Facebook using keywords and social network structure.
3.	G. Patil et. al; 2013	OW	Identifying and blocking terrorist websites using content analysis.
4.	M. GoodWin; 2013	MB, VS, SN	Analysis of various counter-jihad, Islam and Muslim communities on web 2.0.
5.	P. Wadhwa et. al; 2013	MB, VS, SN	Dynamic tracking of radical groups on web 2.0 by analyzing messages and post.
6.	H. Chen et. al ; 2012	DF, VS	Examine several dark web forums and videos used by terrorist & extremist groups.
7.	D. Denning et. al; 2012	VS, SN	An in-depth research on Social Media associated with jihad and counter terrorism.
8.	C. Logan, et. al ; 2012	BL, OW	Finding similarities between different extremist groups using thematic content analysis.
9.	S. Mahmood; 2012	MB, VS, SN	Comparing several defense mechanisms to detect terrorists on social network websites.
10.	J. Hawdon; 2012	MB, VS, SN	A statement about the effect of hate groups as hate-inspired violence on the web.
11.	O'Callaghan et. al; 2012	MB, VS, SN, OW	Activity and links analysis of extreme right groups (local and international) on Twitter.
12.	E. Erez; 2011	DF	Quantitative and qualitative assessments of the content of communications on forums.
13.	D. David et. al; 2011	MB, SN	Detecting criminal groups & most visible players using the keyword search & contacts.
14.	A. Sureka et. al.; 2010	VS	Locating hate promoting videos, users and their groups sharing a common agenda.

2. To investigate the effectiveness of contextual features such as the title of the videos uploaded, commented, shared, and favourited for computing the similarity between nodes in the focused crawler traversal. To examine the effectiveness of subscribers, featured channels and public contacts as links between nodes.

3. To conduct a case-study by defining a specific topic (anti-India) and perform an in-depth empirical analysis on real-world data from YouTube.

4. To conduct a characterization study of the anti-India hate and extremism promoting videos based on terms present in the title of the videos, YouTube category, average length of videos, content focus and target audience

5. To discover user communities and groups and apply Social Network Analysis (SNA) based measures (such as centrality) to identify core users.

2   Related Work and Research Contributions

In this section, we discuss closely related work to the study presented in this paper. We conduct a literature survey on the topic of hate and extremist content detection on Web 2.0. Table 1 shows a list of 14 papers in reverse chronological order. As shown in Table 1, we characterize the papers on the basis of the social media platform and the objective of analysis. Table 1 reveals that researchers have conducted experiments on several social media platforms such as video-sharing website, micro-blogging websites, online discussion forums and social networking websites. Table 1 shows a diverse domain of study covered in existing literature: terrorism, extremist groups, anti-black communities, US domestic, middle eastern, jihad and anti-Islam.

1. I-Hsien Ting et. al. propose an architecture to discover hate groups on Facebook using text mining and social network analysis. Extracted features include keywords that are frequently used in groups [18].
2. M. Goodwin analyses several hate and extremist groups coming into existence across various countries. He presents an in-depth analysis of their activities, supporters and reasons behind the emergence of these groups [9].
3. A. Sureka et. al. propose an approach based upon the data mining and social network analysis in order to discover hate promoting videos, users and their hidden communities on YouTube [17].
4. H. Chen et. al present a framework to identify extremist videos on YouTube. They extracted lexical, syntactic and content specific features from user generated data and applied different feature based classification techniques to classify videos [4] [6] [5] [8].
5. E. Reid et. al present a hyperlink study to discover US extremist groups and their online communities on various discussion forums and video sharing websites. They perform web crawling and text analysis on the web content in order to find the relevant websites [14].
6. A. Salem et. al propose a multimedia and content based analysis approach to detect jihadi extremist videos and the characteristics to identify the message given in the video [15].

In context to existing work, the study presented in this paper makes the following unique contributions (the study presented in this paper is an extension of our previous work [1]):

1. We present an application of focused or topical crawler based approach for locating hate and extremism promoting channels on YouTube. While there has been a lot of work in the area of topical crawling of web-pages, this paper presents the first study on adaptation of focused crawler framework (best-first search and shark-search) for navigating nodes and links on YouTube.
2. We conduct a series of experiments on real-world data downloaded from YouTube to demonstrate the effectiveness of the proposed solution approach by varying several algorithmic parameters such the size of n-gram for language modeling based statistical model, similarity threshold for the text classifier, starting point or seed for best-first search and shark search version of the algorithm.
3. We perform a characterization study of the anti-India hate and extremism promoting videos based on terms present in the title of the videos, YouTube category, average length of videos, content focus and target audience. We apply

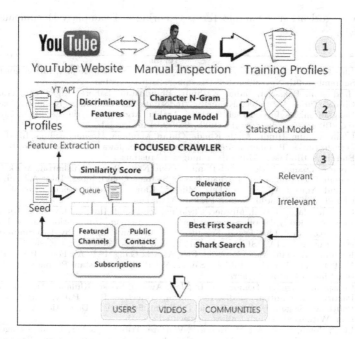

Fig. 1. A General Research Framework For Our Proposed Solution Approach

Social Network Analysis (SNA) based techniques on the retrieved user profiles and their connections obtained from the focused crawler traversal to understand presence of communities and central users.

3   Research Framework and Methodology

Figure 1 presents a general framework for the proposed solution approach. The proposed method is a multi-step process primarily consists of three phases, Training Profile Collection, Statistical Model Building and Focused Crawler cited as Phase 1, 2 and 3 respectively.

We perform a manual analysis and a visual inspection on activity feeds and contextual metadata of various YouTube channels. We collect 35 positive class channels (promoting hate and extremism) used as training profiles. We build our training dataset by extracting the discriminatory features (user activity feeds-titles of videos uploaded, shared, favourited & commented by the user and profile information) of these 35 channels using YouTube API[4]. In the training dataset, we observe several terms relevant to hate and extremism and divide them into 9 main categories shown in Table 2. We build a statistical model from these training profiles by applying character n-gram based language modeling approach.

We chose character-level analysis (low-level features) as it is language independent and does not require extensive language specific pre-processing. The other

---

[4] https://developers.google.com/youtube/getting_started

Table 2. Categorization of Sample Terms Occuring in Examplary Documents for Focused Crawler

Category	Terms
Important Dates	13th January, 26th January, 23rd March, 5th August, 14th August, 15th August, 21st September, 9th November, 3rd December, 25th December
Region	Hindustan, Pakistan, India, Kashmir, Bhindustan, Lahore, Afganistan, America, China, Turkey, Mumbai, Khalistan, Indo-Pak, US, Jammu & Kashmir, Agartala, Bangladesh, England, Israel, Karbala, Arabia, Argentine, Syria, Egyptian, Goa, Orrisa, Bihar, Canadian, Arab, Sindh, Balochistan, Punjab
Religion	Islam, Muslim, Hindu, Allah, Khuda, Quran, Maulana, Mosque, Kabba, Jihad, Azan, Jewish, Burka, Prophet, Religious, Koum, Islamic, Jews Christians, Apostates, Sikh, Buddhist, Hinduism, Muhajirs, Immigrant Muslims
People Name	Obama, Osama, Laden, Zaid Hamid, Zakir Naik, Parvez Musharraf, Mark Glenn, Jinnah, Saed Singh, Imran Khan, Nawaz Shareef, Quaid, Iqbal, Tahir Ashrafi, Emad Khalid, Yousuf Ali, Shaykh Feiz, Mustafa Kamal, Khalid Yaseen, Asma Jahangir, Chandragupt, Gandhi, Nehru, Pramod Mahajan
Negative Emotions	Horrible, Hate, Hatred, Murder, Cheating, Ice-Blood, Honour, Loathing, Humanity, Violence, Bloody, Blood, Revenge, Torture, Extremism, Humiliation, Abuse, Poverty, Fear, Scoundrel, Lies, Fraud, Friendship, Hesitation, Fake, Filthy, Discrimination
Communities	Paki Punjabi, CIA, ISI, Takmel-E-Pakistan, Brass Tacks, Azad Kashmir, Liberate Kashmir, Taliban, Aman Ki Asha, Flag Attack, Gang, IAF, Air Force, RAW, PMLN, NATO, TTP, Threek-E-Taliban, SWAT, WUP, PPP, Pakistani People Party, Operation Shudhi Karan, Aryavrat
Politics Terms	Conspiracy, Leader, Democracy, Inqalab, Awami, Strike, Khilafat, Against, Rights, Partition, Corruption, Media, Resolution, Objective, Rule, Party, League, Protest, Politician, Slogan, Division, Public, President, Secularism, Domestic, Congress, Election, Witnessed, Tribal, Rallies, persecuted, Youth
War Related Terms	LOC, Bomb, Blast, Attack, Holywar, Warfare, Tribute, Soldier, Jawan, Refugee, Enemies, Fighting, Patriot, Assassination, Expose, Propaganda, Army, Protocol, Security, Anthem, Threat, Nukes, Border, Shaheedi, Military, Zindabad, Hijab, Dirty War, Black Day, Terror, Mission, Operation, Jail, Prison, Open Fire, Destruction, Halaal, Grave, Sectarian, Genocide, Encounter, Ghadar, Strategy, Battle Field, Nation, Warning, Killing, Legendary, Campaign, Ghulami, Weapon, Qarz, Unsafe, Insulting, Defend, Accident, Judicial, Failure, Camp, Evil, Vision, Armed Forces, Agent, Martyred, Missing, Intentions, Defeat, Secret, Slap, Traitor, Reclaim, Tragedy, Shahadat, Accusing, TAKBEER, Terrorists, War On Terror, Crime, Bloodshed, Revolution, Constitution, Vandalism, Victorious, Violation, Graves, Torture, Slaughtered, Explodes, Struggle, War, Freedom, Jet Carrier, Police men, Slave, Honour killing
Others	Pig, Monkey, Faith, Ideology, Earthquake, Thunder, Uneducated, Awareness, Debate, Foreign, Leaked, Press, Affair, Economic, Destiny, Flood, Endgame, Rebuttal, Documentary, Respect, Argue, Patrol, Scandal, Survival, Rapist, Rape, Ideological, Geographical, Sections, Sects, Government, Interview

advantage of character n-gram based approach is that it can capture sub-word and super-word features and is suitable for noisy text found in social media. The paper by Peng et al. lists the advantages of character-level n-gram language models for language independent text categorization tasks [12]. In phase 3, we build a focused crawler (best first search and shark search) which is a recursive process. It takes one YouTube channel as a seed (a positive class channel) and extract it's contextual metadata (user activity feeds and profile information) using YouTube API. We find the extent of textual similarity between these metadata and training data by using statistical model (build in phase 2) and LingPipe API[5]. We implement a binary classifier to classify a user channel as relevant or irrelevant. A user channel is said to be relevant (hate and extremism promoting channel) if the computation score is above a predicted threshold. If a channel is relevant, then we further extend it's frontiers (links to other YouTube

---

[5] http://alias-i.com/lingpipe/index.html

channels) i.e. the subscribers of the channel, featured channels suggested by the user and it's contacts available publicly. We extract these frontiers by parsing users' YouTube homepage using jsoup HTML parser library[6]. We execute focused crawler phase for each frontier recursively which results a connected graph, where nodes represent the user channels and edges represent the links between two users. We perform social network analysis on the output graph to locate hidden communities of hate promoting users.

3.1   Solution Implementation

In this section, we present the methodology and solution implementation details for the design and architecture articulated in the previous section. In focused crawler we first classify a seed input as relevant or irrelevant which further leads to more relevant channels. In proposed method we use focused crawler for two different graph traversing algorithms i.e. Best First Search (BFS) Algorithm and Shark Search Algorithm (SSA). Algorithm 1 and Algorithm 2 describe the focused crawlers we develop to locate a group of connected hate and extremist channels on YouTube. The result of both algorithms is turned out to be a directed cyclic graph where each node represents a user channel and an edge represents a link between two users. The goal of BFS and SSA is to first classify a channel to be relevant (positive class) or irrelevant (negative class) and then exploring the frontier channels of a relevant user (in case of BFS) and both users (in case of SSA).

Inputs to these algorithms are a seed (a positive class user) U, width of graph w i.e. maximum number of children of a node, size of graph s i.e. maximum number of nodes in graph, threshold th for classification, n-gram value Ng for similarity computation, and a lexicon of 35 positive class channels U_p. Table 4 shows a list of all seed inputs we have used for different iterations. We compare each training profile with all profiles and compute their similarity score for each mode. We take an average of these 35 scores and compute the threshold values. Both algorithms are different in their approach explained in following subsections:

Focused Crawler- Best First Search. The proposed method (Algorithm 1) follows the standard best first traversing to explore relevant user to seed input. Best-First Search examines a node in the graph and finds the most promising node among it's children to be traversed next [13]. This priority of nodes (users) is decided based upon the extent of similarity with the training profiles. A user with the similarity score above a specified threshold is said to be relevant and allowed to be extended further. If a node is relevant and has the highest priority (similarity score) among all relevant nodes then we extend it first and explore it's links and discard irrelevant nodes. We process each node only once and if a node appears again then we only include the connecting edge in the graph.

Steps 1 and 2 extract all contextual features for 35 training profiles using Algorithm 3 and build a training data set. Algorithm SSA is a recursive function

---

[6] http://jsoup.org/apidocs/

Algorithm 1: Focused Crawler- Best First Search

Data: Seed User U, Width of Graph w, Size of Graph s, Threshold th, N-gram Ng, Positive Class Channels U_p
Result: A connected directed cyclic graph, Nodes=User u

```
1   for all u ∈ U_p do
2   |   D.add(ExtractFeatures(u))
    end
    Algorithm BFS(U)
3       while graphsize < s do
4           userfeeds U_f ←ExtractFeatures(U)
5           score score ←LanguageModeling(D, U_f, N_g)
6           if (score <th) then
7           |   U.class ←Irrelevant
            else
8           |   U.class ←Relevant
9           |   Hashmap U_sorted.InsertionSort(U, score)
            end
            for i ← 1 to w do
10          |   Hashmap U_graph.add(U_sorted(i))
            end
11          for all U_g ∈ U_graph do
12          |   fr = Extract_Frontiers(U_g)
13          |   Hashmap U_crawler.add(fr)
            end
14          for all U_fr ∈ U_crawler do
15          |   BFS(U_fr)
            end
    end
```

---

which takes U as a seed input. Steps 4 and 5 extract all features for seed user U and compute it's similarity score with training profiles using character n-gram and language modeling (using LingPipe API). Steps 6 to 8 represent the classification procedure and labeling of users as relevant or irrelevant depending upon the threshold measures.

BFS method has non-binary priority values assigned to each node. The priority values are the similarity score, which is computed by comparing the users' contextual metadata (user activity feeds and profile information) with training profiles. Steps 9 and 10 make a list of top w (maximum number of children, a node can have) users among relevant users based upon their similarity score, sorted in a decreasing order. Step 16 extracts frontiers of a user channel using Algorithm 4. Steps 18 and 19 repeat steps 3 to 15 for each frontier extracted. We execute this function till we get a graph with desired number of nodes or there is no more node is left to extend.

Focused Crawler- Shark Search. We propose a focused crawler for Shark Search Algorithm (Algorithm 2), an adaptive version of the same algorithm introduced in M. Hersovici et. al. [10]. Shark Search algorithm is different from Best First Search algorithm in a way that it explores frontiers of both relevant and irrelevant nodes. In SSA if the parent of a node is an irrelevant node then the inherited score of the child node is $score_{child} * d$, where d is a decay factor, an extra input for SSA which directly impacts on the priority of user. This inherited score is dynamic because a node can have more than one parent.

Steps 1 to 5 are similar to Best First Search (Algorithm 1). Steps 6 to 9 check if the user is a child of irrelevant node then it computes an inherited score for the user by multiplying the original score by a decay factor d. If a node has appeared before and has not been extended further then we update it's similarity score by the maximum value of old and new inherited score. Steps 10 to 12 represent the

---

Algorithm 2: Focused Crawler- Shark Search

Data: Seed User U, Width of Graph w, Size of Graph s, Threshold th, N-gram Ng, Positive Class Channels U_p, Decay Factor d
Result: A connected directed cyclic graph, Nodes=User u

```
 1  for all u ∈ U_p do
 2  |   D.add(ExtractFeatures(u))
    end
    Algorithm SSA(U)
 3  |   while graphsize < s do
 4  |   |   userfeeds U_f ←ExtractFeatures(U)
 5  |   |   score score ←LanguageModeling(D, U_f, Ng)
 6  |   |   if (U is a child of Irrelevant node) then
 7  |   |   |   score ← score * d
    |   |   end
 8  |   |   if (U has appeared before) then
 9  |   |   |   score ← max(new_score, old_score)
    |   |   end
10  |   |   if (score <th) then
11  |   |   |   U.newclass ←Irrelevant
    |   |   else
12  |   |   |   U.newclass ←Relevant
    |   |   end
13  |   |   Hashmap U_sorted.InsertionSort(U, score)
    |   |   for i ← 1 to w do
14  |   |   |   Hashmap U_graph.add(U_sorted(i))
    |   |   end
15  |   |   for all U_g ∈ U_graph do
16  |   |   |   fr = Extract_Frontiers(U_g)
17  |   |   |   Hashmap U_crawler.add(fr)
    |   |   end
18  |   |   for all U_fr ∈ U_crawler do
19  |   |   |   SSA(U_fr)
    |   |   end
    |   end
```

---

classification procedure and labeling of users as relevant or irrelevant similar to Algorithm 1.

The SSA method also uses non-binary priority values same as similarity score of users. Steps 13 and 14 make a list of top w (maximum number of children, a node can have) users (could be relevant or irrelevant unlike BFS) based upon their similarity score, sorted in a decreasing order. Steps 15 to 19 extract frontiers of a user channel using Algorithm 4 and repeats steps 3 to 19 for each linked user.

Features Extraction. In Algorithm 3, we retrieve contextual metadata of a YouTube user channel using YouTube API. Step 1 extracts the profile summary of the user. Steps 2 to 5 extract the titles of videos uploaded, commented, shared and favourited by given user U. The result of the algorithm is a text file containing all the video titles and user profile information.

Algorithm 3: FEATURES EXTRACTION FOR A YOUTUBE USER

Data: User u
Result: User Activity Feeds and Profile Information

```
    Algorithm ExtractFeatures(U)
 1  |   u_Profile ←u.getSummary()
 2  |   u_Uploads ←u.getUploadedVideo()
 3  |   u_Commented ←u.getCommentedVideo()
 4  |   u_Shared ←u.getSharedVideo()
 5  |   u_favorited ←u.getFavoritedVideo()
```

Algorithm 4: FRONTIER EXTRACTION FOR A YOUTUBE USER

Data: User u
Result: Frontiers of a channel
Algorithm $Extract\_Frontiers(U)$
1      $u_{subs} \leftarrow$ u.getSubscribers()
2      $u_{fc} \leftarrow$ u.getFeaturedChannels()
3      $u_{con} \leftarrow$ u.getFriends()

Table 3. List of Few Users Ids of Hate and Extremism Promoting Videos Being Used As Exemplary Documents For Training A Text Classifier

AabeKosar	BTghazwa	haider2026	IndianVictim
Ahmad12791	charbi88	issabln2011	kashafsha
amiruddinmughal	GobletG	GreaterPakistan	khawajak
azadkashmiriboy	hijazna	HinduismIslam	junihashmi
BrassTacksOfficial	netdarvin	IndiaEternal	GreenEye1947
PakistanKaKhudaHafiz	p4pathanp4pakistan	sabeqoonwaawaloon	TAKMEELEPAKISTAN

Table 4. Name of 10 Seed Inputs Used for BFS and SSA- Row-wise Ordered

TheGreaterPakistan	BTghazwa	GreaterPakistan	PakistanRoxxx	PakistanKaKhudaHafiz
BrassTacksOfficial	haider2026	hiddenpakistani	PakistanHeaven	MujheHayHukmeAzan

Frontiers Extraction. In Algorithm 4, we extract all external links of a YouTube channel to other YouTube channels. These links could be the subscribers, featured channels (suggestions by user) and public contacts (friends). YouTube API does not allow users to retrieve the contacts of other users which is why we use jsoup HTML parser library to fetch all frontiers and public contacts list. This algorithm returns a vector of all channels user U is linked with and we make sure that there is no redundant channel in the list.

4 Empirical Analysis and Performance Evaluation

In this section we present the characterization of hate and extremist videos. We demonstrate the experiments and analysis set up, performance results and the effectiveness of our proposed solution approach.

4.1 Experimental Dataset

Training Dataset. A focused crawler needs to classify if a given web-page is relevant or not with respect to a topic. The crawler requires exemplary documents or training examples to learn the specific characteristics and properties of documents in the training dataset. A statistical model (text classifier) needs to be built from a collection of documents pertaining to a predefined topic. Table 3 shows a list of few user ids (channel names on YouTube) used as a training profiles. These user ids consists of 612 videos and hence the training is performed on 612 videos. We obtain the training dataset by manually searching (keyword

Table 5. Results of Focused Crawler for 6 Different Seeds. Modes Represent 6 Different Thresholds (Th) & N-gram (Ng) Pairs. A: Th=-2.0, Ng=3, B: Th=-2.5, Ng=3, C: Th=-3.0, Ng=3, D: Th=-2.0, Ng=5, E: Th=-2.5, Ng=5, F: Th=-3.0, Ng=5.

(a) Focused Crawler- Best First Search

Seed	Seed 1						Seed 2						Seed 3					
Mode	A	B	C	D	E	F	A	B	C	D	E	F	A	B	C	D	E	F
Relevant	26	19	58	21	56	60	39	57	30	26	64	67	1	1	1	1	1	1
Irrelevant	23	2	9	3	11	7	11	1	4	5	1	1	0	0	0	0	0	0
Processed	119	448	239	134	159	145	119	448	239	134	312	283	1	1	1	1	1	1
Graph	23	19	25	21	26	26	23	24	25	22	26	26	1	1	1	1	1	1
Minimum	-2.13	-2.7	-6.83	-6.3	-4.75	-4.75	-9.43	-9.43	-6.83	-0.84	-8.43	-8.43	-1.78	-1.78	-1.78	-1.01	-1.01	-1.01
Maximum	-2.13	-0.97	-0.52	-0.48	-0.48	-0.48	-0.8	-0.8	-0.52	-0.46	-0.46	-0.46	-1.78	-1.78	-1.78	-1.01	-1.01	-1.01
Median	-2.13	-1.73	-1.91	-1.23	-1.23	-1.23	-2.08	-1.85	-1.72	-1.7	-1.7	-1.7	-1.78	-1.78	-1.78	-1.01	-1.01	-1.01
Quartile 1	-2.13	-2.13	-2.21	-1.39	-1.78	-1.78	-2.26	-2.26	-2.19	-2.16	-2.08	-2.1	-1.78	-1.78	-1.78	-1.01	-1.01	-1.01
Quartile 3	-2.13	-1.3	-1.54	-0.87	-0.87	-0.87	-1.65	-1.58	-1.5	-1.2	-1.2	-1.19	-1.78	-1.78	-1.78	-1.01	-1.01	-1.01

Seed	Seed 6						Seed 7						Seed 8					
Mode	A	B	C	D	E	F	A	B	C	D	E	F	A	B	C	D	E	F
Relevant	5	34	27	32	32	32	0	28	25	21	31	31	1	1	1	1	1	1
Irrelevant	2	4	10	11	6	6	1	5	1	4	2	2	0	0	0	0	0	0
Processed	20	313	290	258	332	332	0	212	318	256	252	274	1	1	1	1	1	1
Graph	5	22	25	22	25	25	0	22	25	21	26	26	1	1	1	1	1	1
Minimum	-2.25	-2.87	-6.83	-6.3	-2.38	-2.38	-2.04	-0.77	-6.83	-4.75	-1.87	-1.87	-1.87	-1.87	-1.87	-1.72	-1.72	-1.72
Maximum	-1.16	-0.97	-0.52	-0.46	-0.46	-0.46	-2.04	-0.97	-0.52	-0.46	-0.46	-0.46	-1.87	-1.87	-1.87	-1.72	-1.72	-1.72
Median	-1.6	-1.68	-1.78	-1.22	-1.29	-1.29	-2.04	-1.77	-1.91	-1.23	-1.33	-1.33	-1.87	-1.87	-1.87	-1.72	-1.72	-1.72
Quartile 1	-1.94	-2.13	-2.21	-1.59	-1.91	-1.91	-2.04	-2.13	-2.24	-1.79	-2.05	-2.05	-1.87	-1.87	-1.87	-1.72	-1.72	-1.72
Quartile 3	-1.34	-1.3	-1.46	-0.82	-0.46	-0.46	-2.04	-1.31	-0.52	-0.89	-1.12	-1.12	-1.87	-1.87	-1.87	-1.72	-1.72	-1.72

(a) Focused Crawler- Best First Search

Seed	Seed 1						Seed 2						Seed 3					
Mode	A	B	C	D	E	F	A	B	C	D	E	F	A	B	C	D	E	F
Relevant	37	34	27	56	29	27	45	29	45	28	29	29	1	1	1	1	1	1
Irrelevant	6	2	2	3	3	2	2	3	0	2	0	0	1	1	1	1	1	1
Processed	198	167	123	177	110	110	138	129	374	122	122	122	0	0	0	0	0	0
Graph	26	26	26	26	26	26	25	26	26	26	26	26	1	1	1	1	1	1
Minimum	-13.9	-13.9	-13.9	-13.2	-132	-132	-2.25	-2.70	-2.70	-2.31	-2.42	-2.42	-1.78	-1.78	-1.78	-1.01	-1.01	-1.01
Maximum	-0.11	-0.12	-0.17	-0.07	-0.15	-0.15	-0.11	-0.13	-1.05	-0.10	-0.46	-0.46	-1.78	-1.78	-1.78	-1.01	-1.01	-1.01
Median	-0.50	-1.46	-1.70	-1.20	-1.21	-1.41	-1.62	-1.47	-1.74	-0.81	-1.18	-1.18	-1.78	-1.78	-1.78	-1.01	-1.01	-1.01
Quartile 1	-1.47	-2.04	-2.26	-1.57	-1.91	-1.97	-1.79	-1.94	-2.30	-1.22	-1.96	-1.96	-1.78	-1.78	-1.78	-1.01	-1.01	-1.01
Quartile 3	-0.21	-0.25	-1.39	-0.85	-0.71	-0.92	-1.26	-1.08	-1.30	-0.23	-0.86	-0.86	-1.78	-1.78	-1.78	-1.01	-1.01	-1.01

Seed	Seed 6						Seed 7						Seed 8					
Mode	A	B	C	D	E	F	A	B	C	D	E	F	A	B	C	D	E	F
Relevant	23	35	27	37	23	23	22	35	26	36	23	23	1	1	1	1	1	1
Irrelevant	4	3	0	0	2	2	2	3	0	2	0	0	0	0	0	0	0	0
Processed	158	169	88	131	80	80	242	213	65	107	54	54	1	1	1	1	1	1
Graph	25	25	25	25	25	25	25	25	25	21	21	21	1	1	1	1	1	1
Minimum	-2.32	-2.70	-2.70	-2.31	-3.62	-3.62	-2.32	-2.70	-2.70	-2.31	-3.62	-3.62	-1.87	-1.87	-1.87	-1.72	-1.72	-1.72
Maximum	-0.05	-0.13	-0.33	-0.07	-0.46	-0.46	-0.05	-0.12	-0.33	-0.17	-0.17	-0.17	-1.87	-1.87	-1.87	-1.72	-1.72	-1.72
Median	-0.23	-1.63	-1.63	-0.87	-1.23	-1.23	-0.20	-1.60	-1.63	-0.97	-1.23	-1.23	-1.87	-1.87	-1.87	-1.72	-1.72	-1.72
Quartile 1	-1.40	-2.12	-2.05	-1.32	-2.05	-2.05	-0.26	-2.25	-1.97	-1.33	-2.05	-2.05	-1.87	-1.87	-1.87	-1.72	-1.72	-1.72
Quartile 3	-0.16	-1.12	-1.60	-0.39	-0.93	-0.93	-0.16	-1.16	-1.63	-0.45	-0.78	-0.78	-1.87	-1.87	-1.87	-1.72	-1.72	-1.72

based) for anti-India hate and extremism promoting channels using YouTube search and traversing related video links (using the heuristic that videos on similar topic will be connected as relevant on YouTube). The training dataset profile consists of profile information of users and the title of videos uploaded, favorited, shared and commented by the user. We believe the title of such videos reflects user interests and can be used for building a predictive model.

Test Dataset. We select 10 random positive class (hate and extremist) channels for creating test dataset. Each user works as a seed input to the focused crawler. Table 4 shows the list of all 10 seeds we select for our experiments. To evaluate the effectiveness of our solution approach we execute our focused crawler sixty times for both Shark Search and Best First Search. Here we use 10 different seeds, 3 different threshold values and 2 different n-gram values for similarity computation. We make 6 pairs of threshold and n-gram values calling them as six different "Modes". For both approaches (BFS and SSA), we run our focused crawler 60 times for 10 seeds and each seed for all 6 modes.

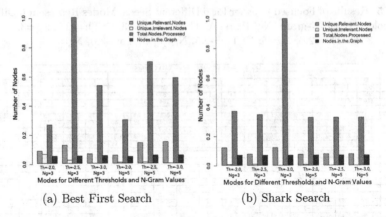

(a) Best First Search   (b) Shark Search

Fig. 2. Illustrating The Variance Between Number of Unique Relevant Nodes, Unique Irrelevant Nodes, Nodes Present in The Graph and Total Number of Nodes Processed for Six Different Modes of Seed 2

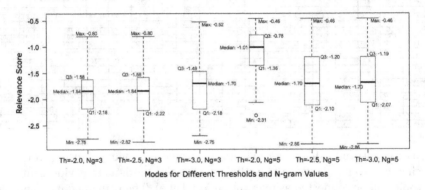

Fig. 3. Box-Plot And Descriptive Statistics For Six Different Configurations Of Best First Search Crawler

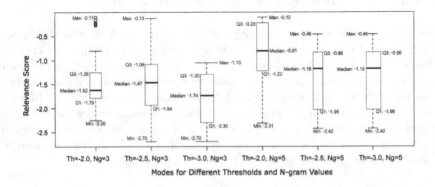

Fig. 4. Box-Plot and Descriptive Statistics for Six Different Configurations of Shark Search Crawler

4.2 Experimental Results

Focused Crawler Results. As mentioned above, we execute our focused crawler 60 times for both BFS and SSA. Table 5 (a) and (b) present the complete picture of users statistics based upon their similarity scores in each iteration. Table 5 (a) and (b) show the number of unique relevant users, unique irrelevant users, total number of users present in the output graph and the total number of users processed during execution of BFS and SSA focused crawlers respectively. Table 5 also shows the summary of similarity scores (minimum, maximum, median, 1st quartile and 3rd quartile) of all users. Table 5 reveals that the number of relevant and irrelevant users vary for different threshold and n-gram pairs. In Table 5 we notice that for both BFS and SSA, five-gram performs better than tri-gram. And for five-gram we achieve maximum number of relevant users in mode F (threshold= -3.0, n-gram= 5). These statistics show that the number of relevant and irrelevant nodes vary for different seeds. For example, for seed 3 and 8 we have only one relevant node. Despite being positive class channels these users have no links to other hate and extremist users on YouTube. Table 5 (a) and (b) reveal the difference in BFS and SSA performance for same seed. For seed 7, 9 and 10, we have an empty graph for BFS while in SSA we have 25 connected users for mode A. And similarly for other modes SSA has more number of relevant users in comparison to BFS.

Figure 2(a) and 2(b) illustrate the variance in number of nodes (shown on Y-axis) for different modes (shown on X-axis) for one seed. Where each node represents a YouTube user. Figure 2(b) depicts that for each mode number of irrelevant nodes for SSA are negligible in comparison to BFS. We also notice that for Seed 2, the graph size is almost similar in both BFS and SSA approach. In BFS we extract frontiers of only relevant nodes unlike SSA. Therefore, for BFS, we see a radical change in number of processed nodes for each mode. For SSA the number of unique relevant nodes as well as the number of processed nodes are similar for all modes except mode C. Figure 3 and 4 show the variance in the statistics of similarity or relevance score (shown on Y-axis) for different modes (shown on the x-axis). These statistics are measured for one seed used for both BFS and SSA approaches and same configuration of threshold and n-gram values. In Figure 3 we see that the first quartile for mode A is below the threshold value and it is smaller than third quartile unlike in Figure 4. It is an evidence that for BFS the number of relevant nodes are lesser in comparison to SSA. In SSA approach we are able to find users which are more relevant (shown as outliers) to training profiles. Figure 3 and 4 show that for modes E and F (Th=-2.5, Ng=5 and Th=3, Ng=5 respectively) all users are classified at relevant.

We asked 3 graduate students of our department to validate our results and they manually annotated each user. Based upon the validation we evaluate the accuracy of our classifier by comparing the predicted class against the actual class of each user channel. Table 6(a) shows the confusion matrix for binary classification performed during Best First Search approach. Given the input of 10 seed users and 6 modes (pair of threshold and n-gram values) we get different

Table 6. Confusion Matrix for Focused Crawlers

(a) Best First Search

		Predicted	
		Relevant	Irrelevant
Actual	Relevant	991	295
	Irrelevant	55	29

(b) Shark Search Algorithm

		Predicted	
		Relevant	Irrelevant
Actual	Relevant	921	314
	Irrelevant	125	67

Table 7. Accuracy Results for Focused Crawler- Best First Search and Shark Search. TPR= True Positive Rate, FPR= False Positive Rate, PPV= Positive Predictive Value, NPV= Negative Predicted Value.

	TPR	TNR	PPV	NPV	F1-Score	Accuracy
BFS	0.75	0.35	0.88	0.18	0.81	0.69
SSA	0.77	0.35	0.95	0.09	0.85	0.74

Table 8. Illustrating The Network Level Measurements for Focused Crawlers- Best First Search (Left) and Shark Search Algorithm (SSA). NN= Number of Nodes, NE= Number of Edges, SL= Number of Self Loops, Dia= Network Diameter, AD= Average Density, ACC= Average Clustering Coefficient, IBC= In- Betweenness Centrality, CC= In- Closeness Centrality, #W/SCC= Number of Weak/Strong Connected Components.

	NN	NE	SL	Dia	AD	ACC	IBC	ICC	#WCC	#SCC
BFS	23	119	3	4	0.225	0.388	0.046	0.356	1	7
SSA	24	137	8	3	0.238	0.788	0.009	0.320	1	16

number of connected users in each iteration. To measure the accuracy of our proposed approach we collect results of all 60 iterations and classify 1046 (921 + 125) users as relevant and 381 (314 + 67) as irrelevant users. There is a misclassification of 25.42% and 65.10% in predicting the relevant and irrelevant users respectively. Table 6(b) shows the confusion matrix for binary classification during Shark Search approach. Given the input of 10 seed users and 6 n-gram & threshold pairs, it classifies 1046 (991 + 55) users as relevant and 324 (295 + 29) as irrelevant users. There is a misclassification of 22.93% and 65.47% in predicting the relevant and irrelevant users respectively. This misclassification occurs because of the noisy data such as lack of information, non-english text and misleading information.

Table 7 shows the accuracy results (precision i.e. PPV, recall i.e. TPR, NPV, TNR, f1-score and accuracy) of focused crawler for both Best First Search and Shark Search approaches. Table 7 reveals that overall SSA approach (accuracy of 74%) performs better than BFS approach (accuracy 69%). Precision and accuracy of SSA are much higher than BFS and similarly recall and f1- score are reasonably higher for SSA.

Social Network Analysis. We perform social network analysis on the output graph of focused crawler, where each node represents a YouTube user channel and each edge represent a relation (friend, subscriber and featured channel) between two users. Table 8 illustrate the network level measurements we perform on the output graphs of BFS and SSA focused crawlers. These values have been computed for seed 2 in mode B (configuration of threshold=-2.5 and n-gram=3). In Table 8 we notice that in SSA approach users are strongly connected in comparison to BFS approach because the average density of network graph is more in SSA approach. Network diameter shows that in SSA each user is reachable in maximum 3 hops while in BFS it takes 4 hops. In SSA, we have more number of connected components than BFS, which helps to locate more communities. Here we see, that SSA has higher clustering coefficient which results into a cluster of highly relevant users.

Figures 5(a) and 5(b) (generated using ORA[7]) show three different representations of network graph, outputs for BFS and SSA focused crawler respectively (seed 2 and mode B- threshold=-2.5, n-gram=3). Graph in the left shows a directed connected cyclic graph. Colors of nodes represent the different in-degree of users and the width of an edge is scaled based upon the number of links between

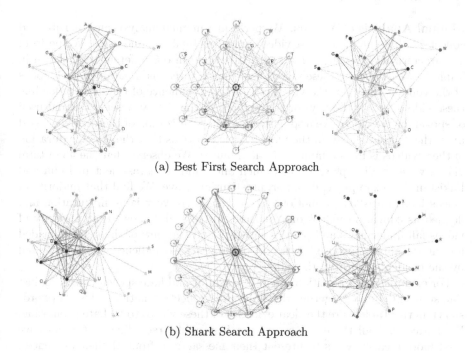

(a) Best First Search Approach

(b) Shark Search Approach

Fig. 5. Community (Left), Betweenness Centrality (Middle) and Cluster (Right) Graph Representation for Best First Search (Top) Shark Search Crawler (Bottom) With Configuration: Th=-2.5, Ng=3 and Seed 2 (Node 'A')

---

[7] http://www.casos.cs.cmu.edu/projects/ora/

two users. In community graphs we see that for BFS all nodes are connected to each other unlike in SSA a few nodes are connected to only one user. Despite the existence of these nodes we find many strongly connected components in SSA which is very less in BFS because all nodes are equally connected. Graphs in the middle of Figures 5(a) and 5(b) are different representation of the output graph based upon the betweenness centrality. Node in the center has the highest centrality among all users and connected to all users of outer shells. In Figures 5(a) and 5(b), graphs in the right are the cluster representation of network. As we see in Table 8, the average clustering coefficient of network in SSA approach is very large in comparison to BFS. Similarly in the Figure 5(a) we see that in BFS approach network has 13 clusters, where total number of nodes is 23. Among these 13 clusters, 6 clusters have only one user node which shows the lack of similarity among users. In Figure 5(b) the cluster representation of network graph (right most graph) has 7 clusters for 24 nodes. where only 2 clusters are formed with one user. In this graph, each cluster shows the level of connectivity to other users. We see the existence of three strong communities made by nodes C, D, E, H, I and B, C, D, E, G, H, I, J and A, D, E, G, where G is the center of all communities and connected to all users.

Manual Analysis of Videos. We perform a manual analysis on YouTube and collect 274 hate and extremist videos uploaded by 35 unique users. We perform a characterization on these 274 and divide them into 5 different sets, shown in Table 9. We categorize these videos based upon three main parameters: 1) focus of the content shown in the videos, 2) targeted audience of the users uploading these videos and 3) the keywords presented in the title & description and used or spoken in the video. We also perform a characterization of these videos based upon the content shown in the video. Table 9 reveals that the average duration of these videos is from 3 minutes to 45 minutes. We observe that the 43 of total videos were small clips showing women and children harassment in India and Pakistan. For example, child labor, prostitution, slave. We find that majority of videos focus on islam promotion. These videos are very large in duration and defined as education videos on YouTube. Table 9 also shows that majority of videos fall under news and politics category and very few of them are uploaded for entertainment purpose. These videos target those audience who are affected by the incidents shown in the videos.

For example, 1947 partition, liberate Kashmir and hate speech videos against Pakistan and India targeting the haters of respective nations. The keywords shown in the Table 9 are the clear evidence of these videos to be hate promoting. We notice that all these videos are not just public recording, but users have used more creative ways to present their messages in front of their audiences. We divide these 274 videos into 12 categories based upon the type of content shown in the video. Table 10 shows that now users have used animation, cartoon, drawings, group discussions and textual messages in their videos to promote hate and extremism. These videos leave a negative impact on the audience and provoke them to write hateful comments.

Table 9. Categorisation of Videos Based Upon Keywords in Video Content & Title and Target Domain of the Uploader, #VD= Number of Videos, YT Category= Youtube Category, Avg Len= Average Duration of Video (in seconds)

#VD	YT Category	Avg Len	Content Focus	Target Audience	Keywords
43	News & Non-Profit	151.68	Honor Killing, Harassment	Women, Refugee People, Child	Honour Killing, Child Marriage, Rape, Responsibility, Protest, Women, Asylum, Arrested, Security, Safety, Refugee, Minor, Brutally, Beating, Exploit, Kidnap, Prostitute, Slave, Indian Police, Delhi, Child Labour.
93	News, Auto, Vehicle, Politics & Education	2526.16	Islam Promotion	Jewish And Muslim People	Vandalism, Jews, Christians, Apostates, Country, Shakyh Abu Hamza, Speech, Hate, Muslim, Fatah Domestic, Leader, Destroy, Killing, Rape, Taliban, Bombs, Battle, Courage, Allah, Islam, Courage, Principle, Politics, Belief, Macca, Trouble, Money, Hatred, Shaheed, Afganistan, Enemies Of Islam, Shame, Rape, Women, Kids, Bad, Women Rights, Debate, Zaid Hamid, Armed Force, Jinnah
25	News, Politics & Education	1225.56	Liberate Kashmir	Kashmiri People	Muslim, Army, Military, 1947, Partition, Azad Kashmir, Liberate Kashmir, Pakistan, India, Killing, Murder, Border, Fighting, Democracy, Martyr, Torture.
83	News & Politics	349.28	Anti-Muslims	Pakistan Haters	Kashmir, Jihad, Pakistan, India, Quran, Muslim, Hindu, Qatil, Zakir Naik, Hate Speech, Masjid, Pandit, Defense, Madarsa, Tribute, Bharat, America, Attack, Napak, Holy, Kabba, Prophet, Strike, Truth, Holy War, Jihad, Al-Queda, Blast, Killed, Enemies Of India, Leftists,Separatists, Maoists, Propaganda, Kasab.
30	Entertainment, Travel, News & Politics	319.61	Anti-India	India Haters	Kashmir, Poverty, Mumbai, Liberate, Hindu, Beggars, Human, Untouchable, Pundit, Casteism, Fraud, Extremism, Attack, Mob, Killed, Anti-Muslim, Anti-Pakistani, Hatred, Masks, Freedom.

Table 10. Categorization of Hate & Extremism Videos Based Upon the Content Shown in the Video

Pictures With Background Music, Animated Videos
Speech, News Segments, Drawing, Interviews , Group Discussion
Lectures, Cartoon And Comics, Debate, Recorded Videos, Textual Messages

5 Conclusions

We present a focused-crawler based approach for identification of hate and extremism promoting videos on YouTube. The accuracy for BFS and SSA versions of the algorithm is 0.69 and 0.74 respectively. Experimental results reveal higher precision, recall and accuracy for shark-search approach in comparison to best-first search. We conduct a series of experiments by varying various algorithmic parameters such as the similarity threshold for the language modeling based text classifier and n-grams. We conclude that by performing social network analysis on network graphs, we are able to locate hidden communities. We identify the users who play major roles in the communities and have highest centrality among all. We reveal the communities by dividing the network graph into clusters formed by similar users. In SSA we find more strongly connected components (16) and communities in comparison to BFS (7).

We perform a characterization on the content and contextual information of several hate promoting videos. The analysis reveals that hate promoting users upload videos targeting some specific audiences. Majority of videos are very large in the duration (3 to 45 minutes). Keywords present in the contextual information and video content are the evidence of these videos doing hate promotion among their viewers.

References

1. Agarwal, S., Sureka, A.: A focused crawler for mining hate and extremism promoting videos on youtube. In: Proceedings of the 25th ACM Conference on Hypertext and Social Media, HT 2014, pp. 294–296. ACM, New York (2014),
   http://doi.acm.org/10.1145/2631775.2631776
2. Agrawal, S., Sureka, A.: Copyright infringement detection of music videos on youtube by mining video and uploader meta-data. In: Bhatnagar, V., Srinivasa, S. (eds.) BDA 2013. LNCS, vol. 8302, pp. 48–67. Springer, Heidelberg (2013),
   http://dx.doi.org/10.1007/978-3-319-03689-2_4
3. Chaudhary, V., Sureka, A.: Contextual feature based one-class classifier approach for detecting video response spam on youtube. In: 2013 Eleventh Annual International Conference on Privacy, Security and Trust (PST), pp. 195–204 (2013)
4. Chen, H.: Extremist youtube videos. In: Dark Web. Integrated Series in Information Systems, vol. 30, pp. 295–318. Springer, New York (2012),
   http://dx.doi.org/10.1007/978-1-4614-1557-2_15
5. Chen, H., Denning, D., Roberts, N., Larson, C.A., Yu, X., Huang, C.-N.: Chapter 1 - revealing the hidden world of the dark web: Social media forums and videos. In: Yang, C., Mao, W., Zheng, X., Wang, H. (eds.) Intelligent Systems for Security Informatics, p. 1. Academic Press, Boston (2013),
   http://www.sciencedirect.com/science/article/pii/B978012404702000001X
6. Chen, H., Denning, D., Roberts, N., Larson, C.A., Yu, X., Huang, C.: The dark web forum portal: From multi-lingual to video. In: ISI, pp. 7–14. IEEE (2011),
   http://dblp.uni-trier.de/db/conf/isi/isi2011.html#ChenDRLYH11

7. Conway, M., McInerney, L.: Jihadi video and auto-radicalisation: Evidence from an exploratory youtube study. In: Ortiz-Arroyo, D., Larsen, H.L., Zeng, D.D., Hicks, D., Wagner, G. (eds.) EuroIsI 2008. LNCS, vol. 5376, pp. 108–118. Springer, Heidelberg (2008), http://dx.doi.org/10.1007/978-3-540-89900-6_13
8. Fu, T., Chen, H.: Knowledge discovery and text mining
9. Goodwin, M.: The Roots of Extremism: The English Defence League and the Counter-Jihad Callenge. Chatham House (2013)
10. Hersovici, M., Jacovi, M., Maarek, Y.S., Pelleg, D., Shtalhaim, M., Ur, S.: The shark-search algorithm. an application: tailored web site mapping. Computer Networks and ISDN Systems 30(1), 317–326 (1998)
11. McNamee, L.G., Peterson, B.L., Peña, J.: A call to educate, participate, invoke and indict: Understanding the communication of online hate groups. Communication Monographs 77(2), 257–280 (2010)
12. Peng, F., Schuurmans, D., Wang, S.: Language and task independent text categorization with simple language models. In: Proceedings of the 2003 Conference of the North American Chapter of the Association for Computational Linguistics on Human Language Technology, vol. 1, pp. 110–117. Association for Computational Linguistics (2003)
13. Rawat, S., Patil, D.R.: Efficient focused crawling based on best first search. In: 2013 IEEE 3rd International Advance Computing Conference (IACC), pp. 908–911 (February 2013)
14. Reid, E., Chen, H.: Internet-savvy us and middle eastern extremist groups. Mobilization: An International Quarterly 12(2), 177–192 (2007)
15. Salem, A., Reid, E., Chen, H.: Content analysis of jihadi extremist groups' videos. In: Mehrotra, S., Zeng, D.D., Chen, H., Thuraisingham, B., Wang, F.-Y. (eds.) ISI 2006. LNCS, vol. 3975, pp. 615–620. Springer, Heidelberg (2006)
16. Sureka, A.: Mining user comment activity for detecting forum spammers in youtube. arXiv preprint arXiv:1103.5044 (2011)
17. Sureka, A., Kumaraguru, P., Goyal, A., Chhabra, S.: Mining youTube to discover extremist videos, users and hidden communities. In: Cheng, P.-J., Kan, M.-Y., Lam, W., Nakov, P. (eds.) AIRS 2010. LNCS, vol. 6458, pp. 13–24. Springer, Heidelberg (2010)
18. Ting, I.-H., Chi, H.-M., Wu, J.-S., Wang, S.-L.: An approach for hate groups detection in facebook. In: Uden, L., Wang, L.S.L., Hong, T.-P., Yang, H.-C., Ting, I.-H. (eds.) The 3rd International Workshop on Intelligent Data Analysis and Management. Springer Proceedings in Complexity, pp. 101–106. Springer, Netherlands (2013), http://dx.doi.org/10.1007/978-94-007-7293-9_11
19. Yin, D., Xue, Z., Hong, L., Davison, B.D., Kontostathis, A., Edwards, L.: Detection of harassment on web 2.0. In: Proceedings of the Content Analysis in the WEB, vol. 2 (2009)
20. Zhou, Y., Reid, E., Qin, J., Chen, H., Lai, G.: Us domestic extremist groups on the web: link and content analysis. IEEE Intelligent Systems 20(5), 44–51 (2005)

ATSOT: Adaptive Traffic Signal Using mOTes

Punam Bedi, Vinita Jindal, Heena Dhankani, and Ruchika Garg

Department of Computer Science, University of Delhi, India
punambedi@ieee.org, vjindal@keshav.du.ac.in,
{heena.cs.du.2013,ruchika.cs.du.2013}@gmail.com

Abstract. This paper presents design and development of Adaptive Traffic Signal using mOTes (ATSOT) system for crossroads to reduce the average waiting time in order to help the commuter drive smoother and faster. Motes are used in the proposed system to collect and store the data. This paper proposes an adaptive algorithm to select green light timings for crossroads in real time environment using clustering algorithm for VANETs. Clustering algorithms are used in VANETs to reduce message transfer, increase the connectivity and provide secure communication among vehicles. Direction and position of vehicles is used in literature for clustering. In this paper, difference in the speed of vehicles is also considered along with direction, node degree, and position to create reasonably stable clusters. A mechanism to check the suitability of cluster initiator is also proposed in the paper. The proposed ATSOT system can be used for hassle free movement of vehicles across the crossroads. Prototype of the system has been designed and developed using open source software tools: MOVE for the Mobility model generation, SUMO for traffic simulation, TraCI for traffic control Interface and Python for client scripting to initiate and control the simulation. Results obtained by simulating ATSOT approach are compared with both OAF algorithm for adaptive traffic signal control and pre-timed approach to show the efficiency in terms of reduced waiting timing at the crossroads. Results are also compared with pre-timed method for single lane and multi-lane environment using Webster's delay function.

Keywords: Motes, VANETs, clustering algorithm, adaptive control, threshold, traffic signals.

1 Introduction

Traffic problem is an extremely significant issue affecting our daily life. One spends a lot of time waiting on the crossroads resulting in significant delay to reach the destination with more fuel consumption. Traffic signals are used to manage traffic on the crossroads. A traffic signal can be pre-timed, semi-actuated or fully-actuated. Pre-timed signal consists of a programmed sequence of various signal phases which is repeated after completion of a cycle. A phase is defined as an assignment of non-conflicting and simultaneous movement of traffic into separate groups to allow collision free vehicle movement [1]. Congestion on crossroads is major disadvantage

W. Chu et al. (Eds.): DNIS 2015, LNCS 8999, pp. 152–171, 2015.
© Springer International Publishing Switzerland 2015

of pre-timed signals when there is no traffic on the road with green light assignment and too much traffic on the road towards red signal. To solve this problem, actuated signals came into the scenario. Actuated signals are capable of altering the phase sequence with varying phase time for a signal depending on vehicle density on the road, which enable them to reduce traffic delays extensively. Actuated signals are further divided into two sub categories: semi-actuated and fully actuated.

Semi-actuated signals use the detectors present only on the major roads, whereas in case of fully actuated signals, all roads (major or minor) have detectors installed on them [1], [2]. The signal is set to provide the green light time as per traffic requirement. The effectiveness of traffic flow across a crossroad varies according to the phases, phase sequences and the timing of the traffic signals installed. Thus, in order to ensure the safety and normal traffic flow at the crossroads, design of adaptive traffic signals becomes essential [3]. The adaptive traffic signals are mainly used for adapting the timings of lights by predicting the traffic volume for smooth traffic flow without congestion.

Most of the cities in the developing countries have pre-time signals, whereas actuated signals are installed on the crossroads in most of the developed countries. There has been significant works done in the area of adaptive traffic signals for congestion free movement of traffic across the crossroads in the literature. There exist a variety of traffic light control systems implemented worldwide like Split Cycle and Offset Optimization Technique (SCOOT) [4], [5], Sydney Coordinated Adaptive Traffic System (SCATS) [6], [7], Real time Hierarchical Optimized Distributed Efficient System (RHODES) [8], Optimization Policies for Adaptive Control (OPAC) [9], Real-Time Traffic Adaptive Control Logic (RTACL) [10] etc. These systems are mainly based on the inductor loop detectors and video sensors installed on the intersections. As per our knowledge, none of these techniques uses clustering mechanism to form stable clusters to adjust the green time for traffic signals.

In this paper, motes are used to sense the traffic flow on the roads which is further used for clustering. Mote is a wireless sensor node, also known as smart dust, gathers sensory information, processes it and then communicates result to other nodes in the network [11]. Mote is a node but vice versa is not true always [12]. VANET is a special type of Mobile Ad hoc Network (MANET) having its root in traffic engineering that permits interaction among the vehicles; vehicle & infrastructure and the infrastructure itself [13], [14]. VANETs are supposed to integrate wireless technology as a type of Wi-Fi [15], Dedicated Short-Range Communication (DSRC) [16], satellite, cellular and WiMAX.

The communication range allowed for VANETs is restricted to 100 m to 300 m. The participating vehicles are turned into routers or wireless nodes for connection and data collection in the allowed range of the network. A new vehicle can join the network when it comes in the range and any vehicle can be dropped out of the network when it goes out of range. VANET can be considered as the backbone for the Intelligent Transportation System (ITS).

VANET is characterised by moving vehicles considered as nodes and fixed roads considered as edges between the nodes to form a graph like structure. There also exist fixed road side units (RSUs) alongside of road to enable the communication among the vehicles. "Fixed infrastructure belongs to the government or private network operators or service providers" [17]. Some of the benefits provided by VANETs are accident prevention, safer roads, congestion reduction and less waiting time. In VANETs messages are broadcasted through a wireless medium and hence communication is at a greater risk. Moreover due to dynamic nature of the vehicles, stale entries as well as congestion exist in the network.

Clustering algorithms ensure security, increase connectivity and reduce overall message transfer in VANETs [18], [19]. The main concern is on finding the stable clusters of vehicles in vehicular network and the job of clustering algorithms is to find not only the minimum number of clusters, but also preserve the existing configuration of the clusters with the least overhead [20]. In VANETs, cluster formation is highly challenging due to highly mobile nodes and unavailability of global topology [21]. In our proposed approach data is collected through motes and is clustered with the clustering approach discussed by Rawashdeh and Mahmud [22]. During the cluster formation process the relative speed of the vehicles along with the position and direction is considered. The clustering based on the above said parameters is found to be stable and efficient.

The clustering algorithm divides the network into clusters wherein highly mobile vehicles are positioned in one cluster and moderately low mobility vehicles are positioned in the alternative group. For the formation of stable clusters, we have devised a mechanism to test the suitability of the cluster initiator. It is required to minimize the cluster initiator transfer. Another consideration for clustering within the VANETs is that the algorithm has to be very fast to minimize time loss. After cluster formation, density is calculated and the information is fed into the proposed adaptive algorithm running on the RSUs. With the received information, signal time is decided according to the algorithm based on threshold. The main idea is to reduce the overall waiting time at the crossroads and ensure speedy movement of the vehicles in order to reduce the fuel consumption.

Rest of the paper is organized as follows. Section 2 reviews the work done in literature in the area of Intelligent Transportation System along with the clustering algorithms used in VANETs. The proposed system ATSOT is presented in section 3. Experimental setup and results are shown in section 4 followed by the conclusion in section 5.

2   Literature Review

Intelligent Transportation Systems (ITS) can be used as a proficient approach to improve the operation of VANETs. Objectives of ITS include comfortable driving, road safety and dissemination of updated road information to the commuters [23].

Adaptive traffic signals have been widely studied in the literature. Examples include the well-known SCOOT, SCATS, RHODES, OPAC, RTACL etc. SCOOT is a centralized traffic responsive system that coordinates the traffic light in a fixed green light sequence as an automatic respond to traffic flow fluctuation in urban areas [4], [5]. SCATS is a system which programs the traffic lights by providing the intelligent pre-defined traffic plans to offer vehicle delay reduction [6], [7]. It relies on loop detectors placed on the lane pavement before the intersection. RHODES uses three level hierarchies for the management of traffic signals [8]. It is designed for under saturated conditions. OPAC is designed for oversaturated conditions [9]. RTACL is designed for networks of streets [10]. RTACL did not meet expected performance measures. OPAC, increased delay and travel time in some instances, and RHODES reduced cycle lengths, but did not show any significant difference in arterial travel times.

Junping Zhang et al. [24] presented a study of multifunctional data driven intelligent transportation system, which can accumulate a massive sum of data collected from diverse resources: Vision-Driven ITS (input data collected through video sensors and used recognition including pedestrian and vehicle detection); Multisource-Driven ITS (e.g. laser radar, inductive-loop detectors and GPS); Learning-Driven ITS (effective prediction of the accidents occurrence to boost the safety of pedestrians by decreasing the effect of vehicle collision);and Visualization-Driven ITS (to assist the decision makers swiftly recognize anomalous traffic patterns and consequently acquire obligatory counteractive actions). In case of complex scenario i.e. with large number of vehicles, there exist some obstacles for object reorganization.

Car-to-car communication based adaptive traffic signal control system is discussed in [25]. The approach reduces the queue length to reduce the waiting time at the crossroads for the vehicles. Clustering is used for the density calculation of approaching vehicles, which is used by the traffic signal controls for setting the cycle timing in this system. DBCV algorithm is used to gather the desired density information by combining cluster and opportunistic dissemination technique. The direction is computed within the vehicle by engaging digital maps and Global Positioning Systems (GPS) in a geographical region and helps in the formation of clusters.

An adaptive traffic light system is designed by Gradinescu et al [26] in which each vehicle is equipped with a short range communication device. The system is centered on a wireless controller node that is positioned on the intersection to determine the optimal values for the signals. Traffic signals can respond to the communication between vehicles to estimate the density of vehicles around it and consequently adjust the signal timing. In [25] and [26] adaptive traffic light systems based on wireless connectivity between fixed controller nodes and vehicles at crossroads are described. These systems are planned and developed with the objective of improving traffic smoothness, reducing the waiting time at crossroads and helping in the prevention of accidents.

Ganesh S. Khekare et al [27] are using the traffic information to avoid accidents. Vehicles themselves are providing this information without the need of extra infra-structure. The authors have developed an adaptive traffic control system similar to the proposal given by [25], [26] in their work. The system is implementing the AODV routing algorithm without considering any clustering method.

In [22], authors have described a speed-overlapped clustering method in highways scenario. In their work, stable and unstable clustering neighbours are defined according to the relative position, movement direction and speed of the vehicles. Clusters are created only among the stable neighbours and clustering can be started only from the slowest or fastest vehicle. Concept of location services is also being used in the earlier work by various researchers in [18], [19], [28], [20], [29]. A problem is also foreseen in method described in [22] if the speed of some vehicle deviates too much from the speed of other vehicles in the cluster, then cluster formation process becomes unstable. But as compared to other methods, this method creates comparatively stable clusters and hence we have used this clustering approach in our work to propose an adaptive algorithm.

Webster's delay function [3] and HCM 2000 [27] are the main delay functions that exist in the literature as per our knowledge. HCM 2000 method is based on the concept of lane groups. Saturation flow and delay are calculated for each lane group separately. In Webster's method, flow is calculated for the total approach width assuming the arrival rate to be random. We have considered the traffic of mixed type with poor lane discipline as present in many developing countries. Hence, Webster's method is used in our work.

Webster's delay model is the standard delay model with the assumption of vehicles arrival at a uniform pace [2], [30], [31], [32], [1], [33]. The expression given by Webster for delay per cycle, d, is as follows:

$$d = \frac{\frac{C}{2}\left[1-\frac{g}{C}\right]^2}{1-\frac{v}{S}} \tag{1}$$

Where g is the effective green time, C is the cycle length, v is the critical flow for the phase and S is the saturation flow. This eq. (1) computes the optimal cycle length that minimizes the average vehicle delay. Effective green time, g, is calculated by the eq. (2) as follows:

$$g = C - t_l \tag{2}$$

Where C is the cycle length and t_l is the lost time. The study done in [34] uses a new job-scheduling based online OAF (oldest arrival first) algorithm in traffic signal control. The results obtained show that the OAF algorithm reduces the delays experienced by the vehicles, as compared with Webster's method and the pre-timed signal control approach.

Literature work focuses on the clustering mechanisms that are mainly based either on direction, speed, position or their combinations. The presented work considers the parameters degree of the node and speed difference along with direction and position while forming the clusters. We also use motes to sense and collect the data that will be fed to the clustering algorithm resulting in the formation of the stable clusters.

Earlier researchers have focused only on the assignment of green time for signals in various phases. Here in this work, optimal green light assignment is done as per traffic on road either by incrementing or decrementing green light timings.

3   Proposed Adaptive Traffic Signal Using mOTes System

The proposed Adaptive Traffic Signal using mOTes (ATSOT) system is based on motes, clustering algorithm and proposed adaptive algorithm. Motes help in collecting data and feed it to the clustering algorithm which forms stable clusters. These stable clusters are used to calculate average queue length which is used by the adaptive algorithm to set the optimal green light timing as per traffic volume on roads for the signal at crossroad.

3.1   Framework of ATSOT

Framework for the proposed ATSOT system is shown in Fig. 1. In ATSOT system, traffic data is sensed and collected through the motes. This data is stored in the databases located locally at RSUs and globally at the traffic centers, and consist of the information about the traffic (e.g. position, density etc. for the moving vehicles). The data collected through motes is fed to the clustering algorithm which runs locally on RSUs and updates the results to traffic centers. For stability of clusters, a formula to check the suitability for becoming cluster initiator is also proposed. From the obtained clusters, average queue length is calculated on that crossroad. Finally, an adaptive algorithm for setting the green light timing is run on the RSUs to set the green light timing for signals at crossroads. RSU simultaneously updates the traffic centers with the results obtained. The algorithm considers minimum threshold, Th_{min} and maximum threshold, Th_{max} during the computation of green light duration for providing congestion free path with least waiting time. Detailed working of the ATSOT system is shown in Fig. 2.

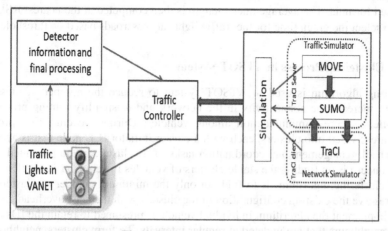

Fig. 1. Framework of ATSOT

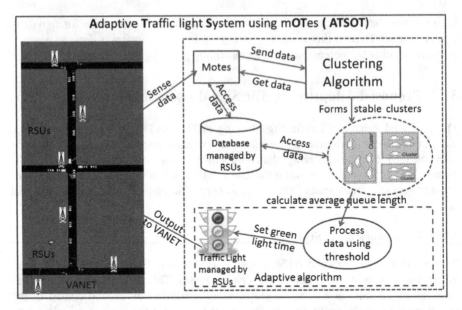

Fig. 2. Detailed working of ATSOT

We assume that each road has a sensor detector installed on the road to calculate the total number of vehicles passing through it. These sensors will transmit the information to the ATSOT system and calculate the cycle timing for next phase.

In Fig. 2, a typical VANET is shown. Motes are installed near traffic lights on the roads. These motes sense the vehicle data and store it in the databases placed at the traffic centers through nearest RSUs. Motes also disseminate the data to the clustering algorithm resulting in formation of the stable clusters of vehicles on the road. These clusters help in computing the average queue length. Clustering algorithm also checks the suitability of the cluster initiator for the formation of stable clusters through the proposed formula. The average queue length is then computed at the nearest RSU and used to select the green time for the traffic lights at crossroads based on threshold.

3.2 Clustering Process in ATSOT System

Clustering algorithm is used in ATSOT system to reduce the number of messages transfer, increase the connectivity in the network and to simplify routing process. It also offers secure communication among vehicles. During the clustering process, nodes are divided into clusters, each with a cluster initiator (CI) node that is accountable for all management and coordination tasks of its cluster. In the presented work, main concern is on finding the stable clusters of vehicles in vehicular network and the job of clustering algorithms is to find not only the minimum number of clusters, but also preserve the existing configuration of the clusters with the least overhead.

To implement the algorithm, individual vehicle is presumed to retain the IDs of the stable neighbours that are updated at regular intervals. To form clusters, neighbouring vehicles cooperate with each other. Vehicles build their neighbourhood relationship using the GPS data. Clusters are formed for the vehicles traveling in the same direction based on their position, speed and node degree. However, the speed among the vehicles varies and this variation might be very high; thus, all neighbouring vehicles are not included in one cluster. Underlying assumption is the existence of a vehicle with the slowest speed among its stable neighbours for initiation of the cluster formation process. The vehicle with slowest speed is called cluster initiating (CI) vehicle.

Since in our work, slower and faster vehicles will be in different clusters, we can start the cluster formation process either from the slowest or fastest vehicle in each direction. The neighbouring vehicles whose relative speed, with respect to the slowest (or fastest) vehicle, is greater than (or less than) the threshold, V_{th}, will not be grouped in the same cluster. In our work, we are choosing the slowest vehicle, V_s to initiate the clustering process by sending a cluster formation request, and then all of its neighbours with relatively similar speed will be in the first cluster. The remaining vehicles will then go through the same cluster formation process to create other clusters. Speed of the vehicle can be extracted from the GPS data. The clustering process starts with the explanation of cluster initiator selection parameters followed by the cluster formation algorithm and cluster maintenance procedure as follows:

Cluster Initiator Selection Parameters

The mobility information (speed, position, degree, and direction) of the nodes is extracted through GPS data. Priority of a node to become a cluster initiator is calculated by its suitability value, u, which is computed based on the mobility information of its neighbourhood.

Thus, u = f (degree, position, speed) is a function defined according to the following criteria:

- The suitability value of the vehicle is calculated by considering the mobility information of vehicle within radius, r and having relatively similar speed only.
- Nodes having higher number of neighbours in their neighbourhood, maintaining closer distances to their neighbours, and having closer speed to the average speed of their stable neighbours should have higher suitability value, thus they are more qualified to be elected as cluster-heads.

Cluster Formation Algorithm

In order to execute the algorithm, each vehicle is assumed to access the GPS data continuously. There exists a vehicle with slowest speed. The algorithm is divided into three phases specified by each of the three algorithms given below.

Algorithm 1 specifies the clustering initiation process followed by the algorithm 2 that explains the process of CI determination with competition among vehicles themselves. Final cluster formation process is described in algorithm 3.

Input : CI ← vehicle ID with slowest speed and stable neighbours.
Output: stable and minimum number of clusters

Algorithm 1: Initiating clustering process

if (there is no vehicle with speed less than that of the current vehicle || vehicle with speed
less than that of the current vehicle ∈ other clusters) then
         temporary CI ← Current vehicle
         send Initiate_Cluster (CI)
end if

Algorithm 2: CI competition and determination

if there are vehicles with speed greater than that of the current vehicle then
         On Receiving Initiate_Cluster (CI)
         CI ← current vehicle
                 check the suitability of the CI w.r.t its neighbouring vehicles having speed
greater than that of the CI
calculate the waiting time before broadcasting admissibility of the current vehicle to
become a CI
while waiting time > 0 do
         if Form_Cluster (CID) is received then
                 if speed of the CI is greater than that CID then
                         Quit_Competition()
                         Process Form_Cluster (CID)
                 end if
         else
                 Decrement waiting time
         end if
         end while
         current vehicle declare itself as Custer initiator with its cluster id as CID
         Send Form_Cluster (CID)
end if

Algorithm 3: Finalizing cluster formation process

if there exist some vehicle with speed difference of V_{th} then
         On Receiving Form_Cluster (CID)
         if that vehicle exists within the radius, r then
                 that vehicle become a Cluster Member
                 cluster id for the vehicle is ID
         else
                 Cluster id will be the default i.e. the vehicle's id
                 Reconstruct the set of the vehicles with speed greater than that of current
vehicle
         end if
end if

Cluster Maintenance

VANETS are highly dynamic in nature. Thus, vehicles keep joining and leaving clusters frequently, causing extra overhead on the maintenance. The maintenance procedure consists of mainly three events that are described as follows:

- **Joining a Cluster:** when a non-clustered vehicle comes within the radius, r of a CI, both the CI and vehicle check whether their relative speeds are within the threshold $\pm V_{th}$ or not. If the speed difference is within $\pm V_{th}$, then that vehicle is added to the cluster members list of that CI. If there is more than one CI in the neighbourhood, then vehicle performs the suitability test to join the cluster.
- **Cluster Merging:** when two CI come within each other's radius r, and their relative speeds are within the predefined threshold $\pm V_{th}$, the cluster merging process takes place. The CI with less number of members gives up its CI role and becomes a cluster-member in the new cluster. The remaining cluster members may join that neighbouring cluster if they exist within the radius of another CI and the speed is also within the threshold. Finally, vehicles that cannot merge with any cluster nor can join a nearby cluster, start clustering process to form a new cluster.
- **Leaving a Cluster:** when a vehicle moves out of the range of cluster radius, r, that vehicle is removed from the cluster members list maintained by the CI. The vehicle changes its state to a standalone if there is no nearby cluster to join or there is no other nearby standalone vehicle to form a new cluster.

3.3   Adaptive Algorithm for Controlling Traffic Lights

After cluster formation, density is calculated and the information is fed into the intelligent traffic signals by the motes. With the received information, signal will be decided according to the proposed adaptive algorithm based on threshold. The main idea is to reduce the overall waiting time at the crossroads and ensure speedy movement of the traffic. The adaptive algorithm is given below:

Adaptive Traffic Control Algorithm

In order to execute the algorithm, clustering process for the vehicles is assumed to be completed. The average queue length is calculated from the clustering information for each direction. Further from the statistical experimentation, the minimum and maximum values for threshold are being selected as Th_{min} and Th_{max} respectively. Output of the algorithm is to find the optimal green light (GL) timing for the crossroad. Default GL time is assumed as 60 seconds in this algorithm which can be chosen as desired. In the proposed algorithm, we are taking T_{min} and T_{max} as the minimum and maximum time needed by the Th_{min} and Th_{max} number of vehicles to pass through the crossroad in one phase respectively. For the green light reduction, we are taking GL equal to T_{min} to pass all Th_{min} vehicles with loss time. Further, for green light extension, we have used an equation to allow maximum number of vehicles to pass in one phase with loss time given as follows:

$$GL = T_{max} + \frac{(n_i - Th_{max})T_{max}}{Th_{max}} \qquad (3)$$

Where GL is the optimal green time, Th_{max} is the maximum threshold, T_{max} is the time taken by Th_{max} vehicles to pass through crossroad in one phase and n_i is the average queue length of the ith phase. The proposed algorithm is based on the following conditions:

- If there are no vehicles in one of the phases that has been allotted the green signal but there are waiting vehicles in the other phase according to information given by detectors, green time extension is allotted to the other phase.
- If there are number of vehicles less than that of the minimum threshold value in one of the phase that has been allotted green light the green light time is set to the minimum time required by signal for passing the crossroad.
- If vehicles are waiting for the green signal at the crossroad on all the phases, then to reduce congestion check if the queued vehicles is greater than some threshold on a phase , green signal is allotted to that phase.
- In the above case when there is queuing on all the phases but queuing is less than threshold then also consider the time that a vehicle has been waiting on the crossroad. If this waiting time is greater than some threshold time limit then change to green the signal to give way to those vehicles.

The parameters used by the algorithm are defined as follows:

L_i : i^{th} phase traffic lights state. In the work, we take 0 for red traffic, 1 for green traffic and 2 for yellow traffic.

P_i : i^{th} phase where i=1, 2, 3, 4.

WT_i : waiting time of vehicle at i^{th} phase.

time_limit: maximum time that a vehicle can wait on the crossroad. This parameter is used to reduce infinite waiting of a vehicle.

Th_{max}, **threshold:** maximum number of vehicles that are queued on each phase.

Th_{min}, **threshold:** minimum number of vehicles that are queued on each phase.

T_{min} : maximum time taken by the Th_{min} vehicles to pass through the crossroad. This parameter can be determined through experiments.

T_{max} : maximum time taken by the Th_{min} vehicles to pass through the crossroad. This parameter can be determined through experiments

Steps of the algorithms are explained as follows:

Algorithm 4: Adaptive traffic control (L, P$_i$, WT$_i$)

While (true)
  for i = 1, 2, 3, 4
    n$_i$=Average Queue Length at P$_i$
    if n$_i$==0
    for all values of j such that i ≠ j
      n$_j$ = Average Queue Length at P$_j$
      WT$_i$=waiting time of vehicle at phase P$_i$
        if n$_j$ == 0
          go to **Label**
        else if WT$_i$ >time_limit || Th$_{min}$ <= n$_j$ <= Th$_{max}$)
            L$_i$=1 with same GL
            go to **Label**
        else if n$_j$ <= Th$_{min}$
            L$_i$=1 with GL = T$_{max}$
            go to **Label**
          else if n$_j$ >= Th$_{max}$
            L$_i$=1 with GL = T$_{max +}$ ((n$_j$ - Th$_{max}$)* T$_{max}$) / Th$_{max}$
    Label: continue the for loop
            endif
    continue the for loop
continue the while loop

---

4   Experimental Setup and Results

The proposed system is simulated using traffic simulators: MObility model generator for VEhicular networks (MOVE) [35], Simulation of Urban MObility (SUMO) [36], [37], [38] and network simulator: Traffic Control Interface (TraCI) [39], [40]. MOVE is a tool that is adopted on the top of SUMO which is open source micro-traffic simulator. MOVE provides a GUI that allows the user to quickly generate real-world simulation scenarios without a simulation scripts. The output of MOVE is mobility trace file that contains real-world vehicle information that can be used by SUMO [41].

SUMO and MOVE are used to model the moving behavior of vehicles and produce traffic information. These two together are traffic simulators and work as server for the system. MOVE is used to simulate a realistic mobility model for VANET in the presented work and its output is given to SUMO which acts as TraCI server. TraCI is the traffic control interface that works as a client for the system and can be programmed using Python scripting language to initiate the communication for further processing in the proposed system. In the simulation, there is interaction with traffic control, which takes the detector information from motes and uses the proposed adaptive algorithm that makes use of threshold to decide the green light time for traffic lights in real time.

The objective of adaptive traffic signals is to minimize the average queue length for the vehicles at the crossroads. In our work we are using motes to sense and collect the data that is fed to the clustering algorithm. For the purpose of analysis we initially design the system to handle single lane traffic, and further extended the simulation for multi-lane traffic. The mobility of vehicles was defined by SUMO that follows the designed road. A random uniform distribution of the speed was specified amongst the vehicles between the range 3 m/s to 9 m/s for the vehicle mobility. The value of distance for the cluster formation is derived from an experimental analysis. The key parameters used for the simulation are summarized in the table 1 below:

To test the accuracy of the system, simulations were run 10 times and the average value was taken to obtain the results. ATSOT is compared with a pre-timed signal system whose cycle time is set by the simulator to be 60 seconds for a phase. This cycle time comprises of the green time and the inter-green time. Further, the measures of validation also included comparison with already existing clustering algorithm. Under the similar simulation conditions we find that ATSOT considerably reduces average queue length experienced by the cars at crossroad. We start the simulation with single lane traffic model with one crossroad as shown below in Fig. 3 and extend the work with multi-lane traffic model with several crossroads as depicted below in Fig. 4.

Table 1. Simulation Setup

Simulation Scenario	Values
Simulation Time	100-4000 s
Vehicle Speed	3 – 9 m/s
Transmission Range	10 m
Number of Simulations for each case	10
Driver Reaction Time	1.5 s
Mobility Model	Car Following
Cycle Time for Pre-timed system	60 s
Length of Car	5 – 8 m

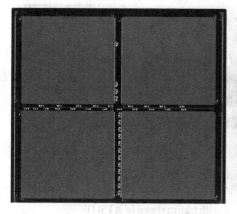

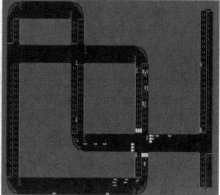

Fig. 3. Screen Shot for single lane traffic   **Fig. 4.** Screen Shot for multi- lane traffic

In single lane traffic model, we consider the straight directions only for moving vehicles. Then extend the simulation with multi-lanes traffic model for vehicle movements in left, straight and right directions. Next, we have used two phases for single lane traffic model as phase 1 and phase 2; and four phases as phase 1, phase 2, phase 3 and phase 4 shown in Fig. 5 for multi-lane traffic models for congestion free movement across the crossroads in the simulation.

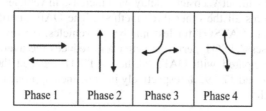

Fig. 5. Different phases used in ATSOT

Fig. 6 depicts the enlarged view of one junction with traffic signals. In the underlying system we developed a simulation in python for pre-timed traffic light control with and without Webster's method under both single lane as well as multi-lane approach. Further we simulated ATSOT system, in which data is collected through motes and fed to the clustering algorithm to form stable clusters. These clusters are used by the adaptive traffic light control algorithm to set the optimal green time for the traffic signals. We also coded the existing OAF algorithm, pre-timed model with and without Webster's delay and finally compare the results in both single lane as well as multi-lane approaches.

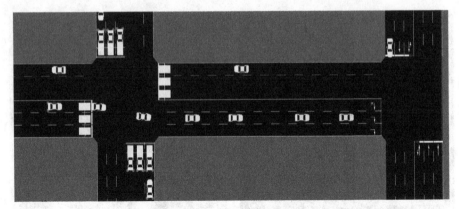

Fig. 6. Enlarged view of a junction with traffic signal in ATSOT

Results are tabulated in Table 2 and Table 3 shown below along with their graphical representation shown in Fig. 7 and Fig. 8. Initially ATSOT system is simulated for single lane approach, and then existing OAF algorithm is also implemented under similar scenario. Further, pre-timed system is simulated under same conditions with and without Webster's model. In the simulation, we gradually increase the number of vehicles and record the average queue length for each case. The results are recorded in Table 2 and their graphical representation is being shown in Fig. 7.

Experimental results show that ATSOT approach implemented in single-lane traffic model, average queue length is less as compared with both OAF and pre -timed approach with and without Webster's delay with increase in vehicles on road. ATSOT approach outperforms all the other three methods. The OAF algorithm performs almost similar as that of ATSOT for less number of vehicles. However, as the number of vehicles increases ATSOT performs better and reduces the average queue length by 1 - 11% as compared with OAF. Further, ATSOT reduced the average queue length by 14 - 83% and 17 - 92% respectively in pre-timed approach with and without Webster's approach.

Table 2. Results for Average Queue Length in ATSOT, OAF algorithm and pre-timed with or without Webster's method in Single lane

Total Vehicles in Simulation	Average Queue Length (ATSOT)	Average Queue Length (OAF algorithm)	Average Queue Length (pre-timed)	Average Queue Length (pre-timed with Webster's)
200	2	2	25	12
400	23	21	50	32
600	74	75	75	75
800	85	95	102	99
1000	139	155	177	169

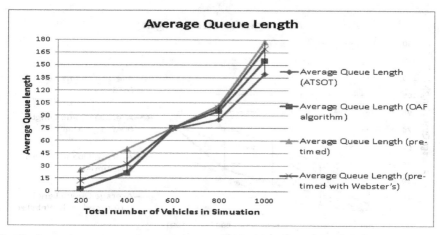

Fig. 7. Graph for Average Queue Length in ATSOT, OAF algorithm and pre-timed with or without Webster's method in Single lane

The simulation was further extended to handle heavy traffic on multi-lane roads. The ATSOT system is implemented for multi-lane approach. The performance of the system is compared with that of existing OAF algorithm under similar conditions. Further, pre-timed system is simulated under same environment with and without Webster's model. In the simulation, we gradually increase the number of vehicles and record the average queue length for each case. The results are recorded in Table 3 and their graphical representation is being shown in Fig. 8.

Experimental results show that ATSOT approach implemented in multi-lane approach, average queue length is less as that of existing OAF algorithm and pre-timed approach with or without Webster as the number of vehicles increases on the road. ATSOT approach outperforms all the other three methods. The ATSOT reduces the average queue length by 15 - 58% as compared with OAF. Further, ATSOT reduced the average queue length by 31 - 78% and 38 - 86% respectively in pre-timed approach with and without Webster's approach.

Table 3. Results for Average Queue Length in ATSOT, OAF algorithm and pre-timed with or without Webster's method in Multi-lane

Total Vehicles in Simulation	Average Queue Length (ATSOT)	Average Queue Length (OAF algorithm)	Average Queue Length (pre-timed)	Average Queue Length (pre-timed with Webster's)
200	3	3	21	10
400	8	19	44	28
600	12	27	62	55
800	34	40	76	62
1000	53	75	86	77

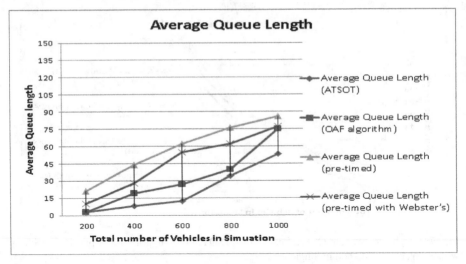

Fig. 8. Graph for Average Queue Length in ATSOT, OAF algorithm and pre-timed with or without Webster's method in Multi-lane

5   Conclusion

Design of Adaptive Traffic Signal using mOTes (ATSOT) system is presented in this paper. The traffic data is sensed through motes and information is collected about the traffic (e.g. position, density etc.) from the moving vehicles. This information is then used by the clustering algorithm to calculate the average queue length. The adaptive algorithm proposed in this paper is based on clustering to find the average queue length. Clustering algorithm also checks suitability of the cluster initiator for formation of stable clusters. Vehicle to vehicle, vehicle to infrastructure and within infrastructure communication is used in this work to set the green light time for crossroads in VANETs. The ATSOT system can help the commuters to get a congestion free path with least waiting time, resulting in overall trip reduction time, dip in the pollution levels, and reduction in fuel consumption. The ATSOT system has been simulated using open source softwares: MOVE, SUMO, TraCI and Python. The results obtained were compared with the pre-timed approach for single lane and multi-lane traffic without and with Webster's delay method and existing OAF algorithm for adaptive traffic signals. Experimental results show that, the proposed ATSOT approach reduces the average waiting time of vehicles at crossroads, as compared with the existing approaches.

Acknowledgement. The authors duly acknowledge the University of Delhi for the support in work on this paper under the research grant number RC/2014/6820.

References

1. Mathew, T.V.: Transportation Engineering I. Mumbai, India: Civil Engineering–Transportation Engineering. IIT Bombay, NPTEL ONLINE (2006)
2. Bell, M.G.H., et al.: Transport Planning and Traffic Engineering. John Wiley & Sons, NY (1997)
3. Webster, F.V.: Traffic signal settings. London: Road Research Laboratory Techn. Rep. No. 39, H.M.S.O (1958)
4. Hunt, P.B., Robertson, D.I., Bretherton, R.D., Winton, R.I.: SCOOT—A Traffic Responsive Method of Coordinating Signals, Transport and Road Research Laboratory, Crowthorne, Berkshire, England, Report TRRL 1014 (1981)
5. Mimbela, L.: Summary of vehicle detection and surveillance technologies use in intelligent transportation systems, Federal Highway Administration, USA, Intelligent Transportation Systems Joint Program Office (2000)
6. Gordon, R.L., et al.: Traffic control systems handbook,Federal Highway Administration, USA, Report FHWA-SA-95032 (1996)
7. Lowrie, P.R.: The Sydney Co-ordinated Adaptive Traffic System—Principles, Methodology, Algorithms. In: Proceedings of the International Conference on Road Traffic Signaling, Institution of Electrical Engineers, vol. 207, vol. 407, London, U.K, pp. 67–70 (1982)
8. Mirchandani, P., Head, L.: A Real-Time Traffic Signal Control System: Architecture, Algorithms, and Analysis. Transportation Research Part C: Emerging Technologies 9(6), 415–432 (2001)
9. Gartner, N.H.: Demand-Responsive Decentralised Urban Traffic Control, U.S.Department of Transportation, Washington, D.C., Report DOT/RSPA/DPB-50/81/24 (1982)
10. Turner-Fairbank Highway Research Center, Adaptive Control Software, Federal Highway Administration., USA, Report HRTS-02-037 (2005)
11. Akyildiz, I.F., Su, W., Sankarasubramaniam, Y., Cayirci, E.: Wireless sensor networks: A survey. Computer Networks 38, 393–422 (2002)
12. Kahn, J.M., Katz, R.H., Pister, K.S.J.: Next century challenges: mobile networking for "Smart Dust". In: MobiCom 1999 Proceedings of the 5th Annual ACM/IEEE International Conference on Mobile Computing and Networking, Washington, USA, pp. 271–278 (1999)
13. Bedi, P., Jindal, V.: Use of Big Data Technology in Vehicular Ad-hoc Networks. In: International Conference on Advances in Computing, Communications and Informatics (ICACCI 2014), Greater Noida, India, pp. 1677–1683 (2014)
14. Mohapatra, P., Krishnamurthy, S.V.: AD HOC Networks: Technologies and Protocols. Springer, Boston (2005)
15. Li, Z., Wang, Z., Chigan, C.: Security of Vehicular Adhoc Networks. In: Wireless Technologies in Intelligent Transportation Systems, pp. 133–174 (2011)
16. Dedicated Short Range Communication (DSRC) Home. (February 2014), http://www.leearmstrong.com/DSRC/DSRCHomeset.htm
17. Car-to-Car Communication Consortium. (April 2014), http://www.car-to-car.org
18. Priyanka, T., Sharma, T.P.: A Survey on Clustering Techniques used in Vehicular Ad Hoc Network. In: Proceedings of 11th IRF International Conference, Pune, Idia, pp. 174–180 (2014)

19. Balia, R.S., Kumar, N., Rodrigues, J.J.P.C.: Clustering in vehicular ad hoc networks: Taxonomy, challenges and solutions. Vehicular Communications 1, 134–152 (2014)
20. Vodopivec, S., Bester, J., Kos, A.: A Survey on Clustering Algorithms for Vehicular Ad-Hoc Networks. In: 35th Iternational Conference on Telecommunications and Signal Processing (TSP), pp. 52–56 (2012)
21. Sood, M., Kanwar, S.: Clustering in MANET and VANET: A Survey. In: International Conference on Circuits, Systems, Communication and Information Technology Applications (CSCITA), Mumbai, India, pp. 375–380 (2014)
22. Rawashdeh, Z.Y., Mahmud, S.M.: A novel algorithm to form stable clusters in vehicular ad hoc networks on highways. EURASIP Jounal on Wireless Communications and Networking 15, 2012 (2012)
23. Shivani, Singh, J.: Route Planning in Vanet By Comparitive Study of Algoriths. International Journal of Advanced Research in Computer Science and Software Engineering 3(7), 682-689 (2013)
24. Zhang, J., et al.: Data-Driven Intelligent Transportation Systems: Survey. IEEE Transactions on Intelligent Transportation Systems 12(4), 1624–1639 (2011)
25. Maslekar, N., Boussedjra, M., Mouzna, J., Labiod, H.: VANET based Adaptive Traffic Signal Control. In: IEEE 73rd Vehicular Technology Conference (VTC Spring), Budapest, pp. 1–5 (2011)
26. Gradinescu, V., Gorgorin, C., Diaconescu, R., Cristea, V., Iftode, L.: Adaptive Traffic Light Using Car-to-Car communications. In: IEEE 65th Vehicular Technology Conference, VTC2007-Spring, Dublin, pp. 21–25 (2007)
27. Khekare, G.S., Sakhare, A.V.: Intelligent Traffic System for VANET: A Survey. International Journal of Advanced Computer Research 2(4(6)), 99–102 (2012)
28. Fan, P., Haran, J.G., Dillenburg, J., Nelson, P.C.: Cluster-Based Framework in Vehicular Ad-Hoc Networks. In: Syrotiuk, V.R., Chávez, E. (eds.) ADHOC-NOW 2005. LNCS, vol. 3738, pp. 32–42. Springer, Heidelberg (2005)
29. Arkian, H.R., Atani, R.E., Pourkhalili, A., Kamali, S.: Cluster-based traffic information generalization in Vehicular Ad-hoc Networks. Vehicular Communications, 11 (2014)
30. Emmelmann, M., Bochow, B., Kellum, C.C.: Vehicular Networking: Automotive Applications and Beyond. John Wiley & Sons Ltd., United Kingdom (2010)
31. Gordon, R.L., Warren, T.: Traffic Control Systems Handbook. Washington, DC: Federal Highway Administration Report FHWA-HOP-06-006 (2005)
32. Roess, R.P., Prassas, E.S., McShane, W.R.: Traffic Engineering, 4th edn. Pearson Higher Education, NY (2011)
33. Wehrle, K., Güne, M., Gross, J.: Modeling and Tools for Network Simulation. Springer, Heidelberg (2010)
34. Kartik Pandit , Dipak Ghosal, H. Michael Zhang, and Chen-Nee Chuah, "Adaptive Traffic Signal Control With Vehicular Ad hoc Networks," IEEE Transactions on Vehicular Technology, vol. 62, no. 4, pp. 1459-1471, May 2013.
35. Lan, K.C., Chou, C.M.: An Open Source Vehicular Network Simulator. In: OSDOC 2011, Lisbon, Portugal, pp. 1–2 (2011)
36. DLR and contributors SUMO-Homepage (2014, October), http://www.dlr.de/ts/en/desktopdefault.aspx/tabid-9883/16931_read-41000/
37. Behrisch, M., Bieker, L., Erdmann, J., Krajzewicz, D.: SUMO - Simulation of Urban MObility - an Overview. In: The Third International Conference on Advances in System Simulation, Barcelona, SIMUL 2011, Spain, pp. 55–60 (2011)

38. Vaza, R.N., Parmar, A.B., Kodinariya, T.M.: Implementing Current Traffic Signal Control Scenario in VANET Using Sumo. In: ETCEE - 2014, National Conference on Emerging Trends in Computer & Electrical Engineering, India, pp. 112–116 (2014)
39. Wegener, A., et al.: TraCI: An Interface for Coupling Road Traffic and Network Simulators. In: Proceedings of the 11th Communications and Networking Simulation Symposium, CNS 2008, Ottawa, Canada, pp. 155–163 (2008)
40. TraCI/Change Traffic Lights State (June 2014), http://sumo.dlr.de/doc/current/docs/userdoc/TraCI/Change_Traffic_Lights_State.html
41. Vaishali, D.: Mobility Models for Vehicular Ad-hoc Network Simulation. International Journal of Computer Applications 11(4), 8–12 (2010)
42. Google's python class - Google for education - Google Developers, https://developers.google.com/edu/python/

Covariance Structure and Systematic Risk of Market Index Portfolio

Lukáš Pichl

International Christian University
Osawa 3-10-2, Mitaka, Tokyo 181-8585, Japan
lukas@icu.ac.jp

Abstract. A set of 18 stocks, selected as the current components of the Dow Jones Index, for which the historical daily closing data quoted at the US market are available for over four decades, is studied. Within this portfolio, we construct a market index with static weights, defined as the relative aggregate trading amounts for each stock. This market portfolio is studied by means of correlation and covariance analysis for the times series of logarithmic returns. Although no measure defined at the correlation/covariance matrices could be found as a definite precursor of market crashes and bubbles, which thus appear as a rather sudden phenomenon, there is an increase in the covariance measures for large absolute values of the logarithmic return of the index. This effect is stronger for the negative values of the log return, corresponding to the market crash case, during which the first principal component of the covariance matrix tends to describe larger proportion of the total market volatility. Periods of low volatility in the market can be characterized by rather significant spread of the relative importance of the first principal component. This finding is common also for the case of dynamically constructed market index, for which the weights are computed as the coordinates of the first principal component eigenvector using short-term covariance matrices.

1 Introduction

Diversification of assets over sectors with negative correlations represents a way of portfolio risk reduction. Taking a financial market as an example, there is, however, a limit that can not be overcome, the systematic risk of the market itself. This is important at the times of turbulence, when the structure of the temporal correlation matrix changes, and all assets exhibit a uniform, upward or downward trend. The latter case is more common, in which the panic drives the entire market to the crash, and the mere fact of exposure to the market, regardless of the particular portfolio composition, ultimately results in large losses. It is an interesting question whether these situations can be predicted based on the market indicators alone, such as the set of all stock prices and the trading volumes. For that purpose, we have selected a set of 18 stocks from the Dow Jones Index and followed them back to the past for the period of

W. Chu et al. (Eds.): DNIS 2015, LNCS 8999, pp. 172–179, 2015.

four decades. Within this limited portfolio universe, we study the dynamics of correlations among the stocks, focusing especially on the covariance matrix. These data are certainly different from the Dow Jones Industrial Average Index (DJIA), since the particular stocks were selected so that no index reconstitution takes place. This procedure can be understood as a sort of specification of static restrains for investment portfolio. In effect, we obtain four decades long time series starting from 1974, which include not only the daily closing prices, but also the daily trading volumes in the units of shares, all available at the site of Yahoo Finance. The first question we followed was whether there exists any precursor to the times of turbulence based on the values of the short-term correlation and covariance matrices. Using scalar products of vectors of logarithmic returns for individual stocks, cosine similarities among returns of shares and indices, Frobenius norm of the covariance matrix, sum of the off-diagonal correlation matrix coefficients, etc., resulted in no significant precursor that could be used to predict large systematic risk, characterized by large absolute values of the logarithmic return of the market index. In other words, the periods of market turbulence and the correlation collapse appear practically simultaneously, as far as the time scale of daily price changes is concerned for the present dataset. We therefore focus on the statistics of covariance matrix features, in relation to the distribution of the logarithmic returns of the market index.

Financial risk and turbulence effects in complex systems [1, 2] have been thoroughly studied both in the fields of econometrics and econophysics. Cross-correlations among assets are subject to time dynamics and exhibit a variety of stylized facts [3]. Among the number of econometric approaches developed for the study of multivariate time series [4], cross-correlations and their dynamics constitute the first line of research [5–8]. For instance, Preiss et al. found an intriguing scaling law for the relation of the sum of the off-diagonal correlation matrix elements and the normalized returns of DJIA on the time scales between 10 and 60 trading days. Persistence of the collective trends in stock market is studied in [9], and the general framework of cross-correlations is complemented by a number of recent studies applied to the particular markets [10]. Although we present the matrix of correlation coefficients for the present dataset in Fig. 1, most of the present study is based on the covariance matrix for the logarithmic returns of individual stock components. Because of the use of the logarithmic returns, this matrix is also dimensionless, albeit not normalized within the range of -1 and 1. The Principal Component Analysis (PCA [11]) is applied to the covariance matrix. This is because the variable with the highest variance in the logarithmic return will dominate the first principal component, which we prefer to the standardization of pairwise correlations. In particular, we study the share of the total variance explained by the first eigenvalue to the total variance within the dataset, thus standardizing our analysis within the range of 0 and 1.

The rest of the paper is organized as follows. In Section 2, the basic variables are defined and the correlation properties of the present dataset pictured. In Section 3, we investigate the properties of the covariance matrix. This is done using first the sum of the off-diagonal covariance elements of the logarithmic returns for individual stock pairs, and then by using the share of the total covariance matrix explained by the first eigenvalue (PCA analysis). The implications of the results on the inner market portfolio dimension are discussed in Sec. 4, and we conclude with summarizing remarks in Sec. 5.

2   Construction of Market Index

The 18 components of the portfolio universe are listed in Table 1 by their commonly used trading name abbreviation. Figure 1 shows the correlation coefficients for the 18 stocks listed in Table 1. This is the standard Pearson's correlation coefficient for the time series of logarithmic returns,

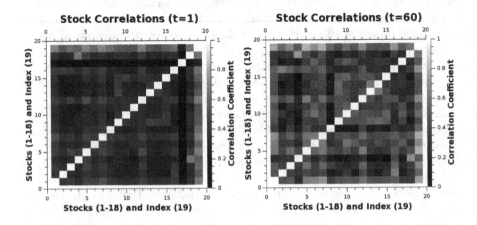

(a) Correlation coefficient on daily scale   (b) Correlation matrix for 60 trading days

Fig. 1. Correlation matrix image for selected portfolio of 18 stocks listed in DJIA. Portfolio index is constructed and labeled as the component No. 19. The correlation coefficients may vary with the time scale step size and the length of the time series (10,290 days between 1974-01-02 and 2014-10-13).

Table 1. Stock specification by trading name abbreviation

AXP (1), BA, CAT, CVX, DD, DIS, GE, IBM, JNJ, KO, MCD, MMM, MRK, PFE, PG, UTX, WMT, XOM (18)

$$R_i(t) = \log \frac{P_i(t+1)}{P_i(t)}, \quad i = 1, \ldots, 18, \quad t = 1, \ldots T. \qquad (1)$$

The label 19 in Fig. 1 is reserved for the market portfolio index, weights of which are defined by the individual share trading amounts $M_i(t)$ (in USD),

$$w_i = \frac{\sum_{t=1}^{T} M_i(t)}{\sum_{i=1}^{18} \sum_{t=1}^{T} M_i(t)}, \quad T = 10,290. \tag{2}$$

The weights are normalized and all positive. Since the matrices of the correlation coefficients include only positive values, an implicit market index derived from the first principal component of the correlation matrix could be used instead. It is known that for the positive correlation matrix, all coordinates of the first principal component vector can be selected with the positive sign. In practice, the role of either index is sufficiently representative, and we thus work with the one given in Eq. (3). The weights are summarized in Table 2.   Figure 2

Table 2. Index weights for 18 stocks in Table 1

w[01]=0.034914, w[02]=0.026008, w[03]=0.032335, w[04]=0.041831,
w[05]=0.022976, w[06]=0.046245, w[07]=0.129413, w[08]=0.096037,
w[09]=0.059690, w[10]=0.076222, w[11]=0.035123, w[12]=0.027158,
w[13]=0.061539, w[14]=0.085853, w[15]=0.056945, w[16]=0.028327,
w[17]=0.048806, w[18]=0.090579

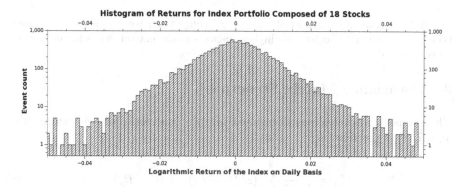

Fig. 2. Statistical distribution of logarithmic returns for the portfolio index constructed from 18 stock components. The weights are static, set as the relative trading amount for each component throughout the entire 40-year long trading period. Logarithmic scale is applied for the relative frequencies to illustrate the normality hypothesis for low values of R; fat tails of the distribution can be observed at both ends of the graph, which are known to exhibit slow, power-law decay.

shows the histogram of the dimensionless logarithmic returns for the static index constructed as described above. The index is determined as the weighted sum,

$$I(t) = \sum_{i=1}^{N} w_i P_i(t), \quad N = 18, \quad t = 1, \dots T. \tag{3}$$

The distribution of trading volumes in currency units for the market is shown in Fig. 3. Since the source data from Yahoo Finance report stock volumes as the number of shares traded, this value is multiplied by the average of the opening and closing prices for each stock, which is accurate enough for the determination of relative trading amounts that serve as the index weights.

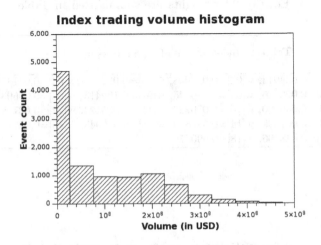

Fig. 3. Histogram of index portfolio trades. Although lower volumes are more frequent, there is a systematic bias towards higher amounts as the market prices inflate over the 40-year period.

3   Covariance Matrix Structure

The sample covariance matrix element for two time series of log returns $R_i(t)$ and $R_j(t)$ is given as

$$C_{ij} = \frac{1}{T-1} \sum_{t=1}^{T} \left(R_i(t) - \bar{R}_i \right) \left(R_j(t) - \bar{R}_j \right), \tag{4}$$

where $\bar{R}_i$, $\bar{R}_j$ are the sample means computed for $t = 1, \dots, T$. We define two covariance measures, $C_{off} = \sum_{i>j} C_{ij}/(N(N-1))$, and C, which is the plain sum of all elements of the covariance matrix. Figure 4 shows the relation of

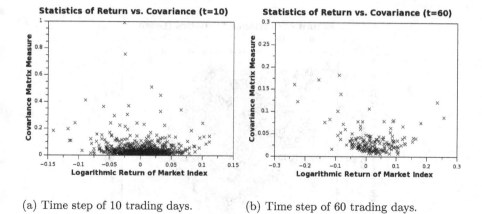

(a) Time step of 10 trading days.   (b) Time step of 60 trading days.

Fig. 4. Off-diagonal aggregate covariance for the logarithmic returns of all 18 index components. The relation to the market index volatility (as measured by the aggregate daily logarithmic return of the index) is shown for two time scales. For $\Delta t = 60$, larger values of logarithmic return of the index correlate to the larger covariance measure (sum of the off-diagonal elements of the covariance matrix).

the log returns of the market index to the covariance measure C_{off}. Two time scales are considered, 10 days and 60 days. In the non-overlapping intervals $I_j = \langle j\Delta T, (j+1)\Delta T - 1\rangle$, the covariance matrices are computed, C_{off} determined, and compared to the market index return, $R_j = \log[I((j+1)\Delta T - 1)/I(j\Delta T)]$, which is the same as the sum of the daily index returns $\log[I(t+1)/I(t)]$ throughout this time interval. The off-diagonal part of the covariance, for low values of R, clusters at the bottom of the figure, although highly correlated events can also be seen at the upper part. These correspond to large volatility events, when the stocks move in unison, but the rises and falls of the index compensate each other. At larger values of R, nevertheless, the covariance measure tends to increase, which is common to both time scales indicated in Fig. 4. We do not observe, nevertheless, any stylized fact that would reduce the cluster onto a single line. The results are similar if the second covariance measure, C, is employed (not shown in the Figure).

4   Market Portfolio Dimension

In order to examine the structure of the covariance matrix more thoroughly, we apply the Principal Component Analysis, PCA. Since C_{ij} is a positive semidefinite matrix, the total volatility of the dataset can be decomposed using the positive eigenvalues λ_i and the orthogonal eigenvectors $v^{(i)}$ of the covariance matrix. Selecting the first principal component, we study the share of the dataset variance it comprises, i.e. the relative measure

$$\rho = \frac{\lambda_1}{\sum_{i=1}^{N} \lambda_i}, \quad \lambda_1 \geq \lambda_2 \geq \dots, \lambda_N, \quad (N = 18). \tag{5}$$

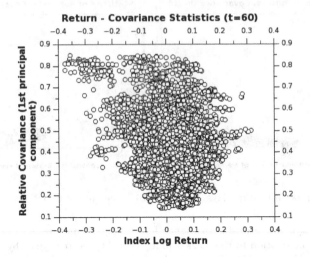

Fig. 5. Statistical distribution of index returns and implied market dimension as measured by the share of the first principal component in the covariance matrix on the total variance of the entire data set. Notice that the quite uniform population for low values of the logarithmic return changes as the return values grow. For larger absolute values of R the first principal component explains a larger share of the market covariance matrix, which corresponds to the collapse of inner market dimension and highly correlated price movement during a period of a bubble or a crash. The trend is more pronounced in market crash situations as shown in the left upper part of the graph. Unlike from Fig. 4, in which the samples do not overlap, moving window on the time series is used here to increase the size of the statistics.

For large values of ρ, near the upper bound of 1, the log returns of all the stocks move in unison, and the intrinsic market dimension reduces as the systematic risk prevails and the price/return movements become uniform. Figure 5 indeed shows this regime is more common when the market index return R is large. Similar to the case of Fig. 4, a significant data spread exists, nevertheless, for low values of R. There is an asymmetry in Fig. 5, namely the values near $\rho = 1$ are more populated for large negative values of R, showing the prevalence of correlated movements in the case of market panic as compared to the case of market euphoria. Similar findings have been found previously in the literature [5]. These findings roughly persist when we alter the definition of the market index, be it an index of prices or index of returns, with the weights defined using the coordinates of the first principal component eigenvector.

5   Concluding Remarks

We have studied a model financial portfolio universe of 18 major stock titles for which (1) 40-year trading history is available and (2) which are at present the formal constituents of the DJIA. Stylized features in the dataset were observed, such as the fat tails in the histogram of the log returns of market index, clustering of the volatility, or the correlation of high covariance measures with large negative values of the index log return. Using the covariance matrix no precursor of market turbulence could be found. Statistical correlation between the situations when the market moves in unison and the aggregate index returns are high was observed, but the emergence of universal features and scaling laws for various-length subsamples of the time series remains to be elucidated.

References

1. Bouchaud, J.P., Potters, M.: Theory of Financial Risks: From Statistical Physics to Risk Management. Cambridge University Press, Cambridge (2000)
2. Sornette, D.: Why Stock Markets Crash: Critical Events in Complex Financial Systems. Princeton University Press, Princeton (2002)
3. Plerou, V., Gopikrishnan, P., Rosenow, B., Amaral, L., Stanley, H.E.: Universal and Nonuniversal Properties of Cross Correlations in Financial Time Series. Phys. Rev. Lett. 83, 1471–1474 (1999)
4. Chib, S., Nardari, F., Shephard, N.: Analysis of High Dimensional Multivariate Stochastic Volatility Models. Journal of Econometrics 134, 341–371 (2006)
5. Preiss, T., Kenett, D.Y., Stanley, H.E., Helbing, D., Ben-Jacob, E.: Quantifying the Behavior of Stock Correlations Under Market Stress. Scientific Reports 2, 752, 1–5 (2012)
6. Conlon, T., Ruskin, H.J., Crane, M.: Cross-correlation Dynamics in Financial Time Series. Physica A 388, 705–714 (2009)
7. Hamao, Y., Masulis, R.W., Ng, V.: Correlations in Price Changes and Volatility Across International Stock Markets. Rev. Financ. Stud. 3, 281–307 (1990)
8. Fenn, D.J., Porter, M.A., Williams, S., McDonald, M., Johnson, N.F., et al.: Temporal Evolution of Financial-Market Correlations. Phys. Rev. E 84, 026109, 1–13 (2011)
9. Balogh, E., Simonsen, I., Nagy, B., Neda, Z.: Persistent Collective Trend in Stock Markets. Phys. Rev. E 82, 066113, 1–9 (2010)
10. Ren, F., Zhou, W.-X.: Dynamic Evolution of Cross-Correlations in the Chinese Stock Market. PLoS ONE 9, e97711, 1–15 (2014)
11. Kritzman, M., Li, Y.Z., Page, S., Rigobon, R.: Principal Components as a Measure of Systemic Risk. J. Portf. Manag. 37, 112–126 (2011)

Moving from Relational Data Storage to Decentralized Structured Storage System

Upaang Saxena, Shelly Sachdeva, and Shivani Batra

Department of Computer Science Engineering
Jaypee Institute of Information Technology, Noida, India
{upaangsaxena,ms.shivani.batra}@gmail.com,
shelly.sachdeva@jiit.ac.in

Abstract. The utmost requirement of any successful application in today's environment is to extract the desired piece of information from its Big Data with a very high speed. When Big Data is managed via traditional approach of relational model, accessing speed is compromised. Moreover, relational data model is not flexible enough to handle big data use cases that contains a mixture of structured, semi-structured, and unstructured data. Thus, there is a requirement for organizing data beyond relational model in a manner which facilitates high availability of any type of data instantly. Current research is a step towards moving relational data storage (PostgreSQL) to decentralized structured storage system (Cassandra), for achieving high availability demand of users for any type of data (structured and unstructured) with zero fault tolerance. For reducing the migration cost, the research focuses on reducing the storage requirement by efficiently compressing the source database before moving it to Cassandra.

Experiment has been conducted to explore the effectiveness of migration from PostgreSQL database to Cassandra. A sample data set varying from 5,000 to 50,000 records has been considered for comparing time taken during selection, insertion, deletion, and searching of records in relational database and Cassandra. The current study found that Cassandra proves to be a better choice for select, insert, and delete operations. The queries involving the join operation in relational database are time consuming and costly. Cassandra proves to be search efficient in such cases, as it stores the nodes together in alphabetical order, and uses split function.

1 Introduction

With growing technology, data and its users are growing exponentially. This exponentially growing data is termed as Big Data. It analyses both structured and unstructured data. Big Data is as important to business and society as the Internet because of the fact that more data leads to more accurate analysis. Thus, data needs to be managed carefully for getting efficient results. Big Data is so large that it is difficult to process using traditional database and software techniques. In most enterprise scenarios, the data is big, moves very fast, and exceeds current processing capacity. Big Data is characterized by five parameters, namely, Volume, Variety, Velocity, Variability and Complexity. 'Volume' refers to the size of the data which determines

W. Chu et al. (Eds.): DNIS 2015, LNCS 8999, pp. 180–194, 2015.

the value and potential of the data under consideration. 'Variety' signifies heterogeneity of Big Data. Big Data consists of not only structured data but also unstructured data. It may constitute images, text, notes, graphs, numbers and dates. 'Velocity' in the context refers to the speed of generation of data or how fast the data is generated and processed to meet the demands and the challenges which lie ahead in the path of growth and development. 'Variability' mentions the inconsistency which can be shown by the data at times, thus hampering the process of being able to handle and manage the data effectively.

Data management becomes very complex process, especially when large volumes of data come from multiple sources. The data needs to be linked, connected and correlated for capturing the information that is conveyed. This situation is termed as the 'Complexity' of Big Data.

The choice of traditional approach (relational model) is assumed to be most promising for storing data due to its power of querying database in an efficient manner rapidly. This assumption proves invalid as data grows in size. Relational databases are not adequate to support large-scale systems due to limitations in their architecture, data model, scalability and performance [1]. This laid down the need for another model which should fulfill the requirement of high availability of data rapidly.

Companies such as Google and Amazon were pioneers to hit problems of scalability and came up with solutions, namely, Big Table [2] and Dynamo [3] respectively. Big Table and Dynamo relax the guarantees provided by the relational data model to achieve higher scalability. Subsequently, a new class of storage systems was proposed named as 'NoSQL' systems. The name first meant 'do not use SQL if you want to scale' and later it was redefined to 'not only SQL' (which means there exist other solutions in addition to SQL-based solutions). NoSQL database named as Cassandra [4] has been proposed by Facebook using the properties of Big Table and Dynamo DB. Cassandra is an open source database and is used to store chats in Facebook [4].

Current research focuses on Cassandra for improving the availability of Big Data. It started with the aim of migrating existing data in PostgreSQL to Cassandra to achieve the benefits of big data analytics. To mitigate overall cost of moving from relational to NoSQL, authors in the current research is performing compression of data before migration using a well-known compression technique, i.e., Snappy Algorithm.

The paper is further divided into following sections. Section 2 highlights the motivation behind the current research and related work. Section 3 proposes a framework for migrating EAV database to Cassandra and provides its implementation details. Section 4 gives the details about experiment performed. Results of the framework implementation are shown in Section 5. Section 6 finally concludes current research and throws a light on future aspects of the work done.

2 Motivation and Related Work

The data is growing exponentially in the world, and we need a special database to handle it. For reducing space, data must be compressed before storage. Many companies are facing the problem of growing data and various methods have been evolved and implemented to provide a commendable solution. Key motivations for the current research are as follows:

- Location dependency due to master-slave behavior of traditional SQL systems.
- Manual intervention during failover and failback situations with generally replicated SQL systems.
- Presence of mixture of structured, semi-structured, and unstructured data.
- Latency and transactional response time issues due to dependence on synchronous replication.
- Costly JOIN operations.
- Absence of method for obtaining sequential information in case of sorted data.
- Low fault tolerance of SQL databases.
- Dynamically allocation of variable length data in database.
- Unavailability of variable schema.

Due to master-slave behavior (centralized storage), availability of data is affected a lot. For improving this, we should move towards distributed storage approach. Ensuing distributed approach helps in various ways as discussed below:

- The use of distributed Database Management System (DBMS) guarantees ZERO downtime (as same data is replicated over multiple nodes and failure of a single node does not impact overall data availability). Whereas, in centralized DBMS, failure of an instance may bring down all the dependent downstream applications. The recovery of centralized DBMS is difficult task, especially when this downtime is not planned.
- Distributed model of DBMS helps in better load balancing across different nodes. Every node is responsible for providing/processing the request for a pre-defined subset of data, which can be configured in central (or master) node, which decides which node will be responsible for providing/ processing which set of data. Whereas, in centralized database there is no secondary instance or replicas available, all the requests need to be processed by central instance. Thus, on a busy day user queries can overwhelm the centralized system, bringing down the responsiveness of the server drastically.
- One of the main advantages of using distributed database is scalability. For increasing the size of database, simply adding one physical instance will work for enhancing the capabilities of database. Whereas in centralized database, it is limited by memory or processing speed of the server where centralized database is mounted on and it cannot be expanded after a certain limit.
- Disaster recovery is easy and reliable in distributed DBMS. Suppose an application is relying on a database which has been mounted on four physical instances (i.e., data has been replicated on four nodes). If one of the nodes fails, recovery will comprise of following steps: (i) Identifying the failure of a node, (ii) Letting the master node (responsible for allocating requests among different nodes), (iii) Removing the failed node from cluster, (iv) Bringing one fresh node / repaired node to the cluster, and replicating the current data from other nodes to this node.

All this happens without impacting applications and users that are sending requests to database. There is no failure at any given point of time for users and external applications. This type of recovery mechanism is not possible in case of centralized DBMS.

Wang G. et. al. reveals the secret of NoSQL in [5]. CAP theorem [6], BASE theorem [7] and Eventual Consistency theorem [8] construct the foundation stone of NoSQL. It is often used to describe a class of non-relational databases that scale horizontally to very large data sets, but in general do not make ACID guarantees. NoSQL data stores vary ideally in their offerings. Consistency means that all copies of data in the system appear same to the outside observer at all times. Availability means that the system as a whole continues to operate inspite of node failure [9]. For example, the hard drive in a server may fail. Partition-tolerance requires that the system continue to operate inspite of arbitrary message loss. Such an event may be caused by a crashed router or broken network link which prevents communication between groups of nodes. Depending on the intended usage, the user of Cassandra can opt for Availability + Partition tolerance or Consistency + Partition tolerance.

To meet the needs of reliability and scalability as described above, Facebook has developed Cassandra [4]. It is one of NoSQL databases used by Twitter, and Facebook. Cassandra is an open source database management system. It is a distributed storage system for managing very large amounts of structured data spread out across many commodity servers, while providing highly available service with no single point of failure [4]. It guarantees the availability of database with Zero downtime at the cost of data redundancy. It aims to run on the top of an infrastructure of hundreds of nodes (possibly spread across different data centers). At this scale, small and large components fail continuously. The way Cassandra manages the persistent state in the face of these failures drives the reliability and scalability of the software systems relying on this service. It resembles database in many ways and shares many design and implementation strategies. However, it does not support a full relational data model, but provides clients with a simple data model that supports dynamic control over data layout and format.

DataStax Corporation examines the why's and how's of migrating from Oracle's MySQL to Cassandra technology [1]. Database migrations in particular can be resource intensive. Thus, IT professionals should ensure that they are taking the right decision before making such a move. Because of rapid expansion of big data applications, various IT organizations are migrating away from Oracle's MySQL or planning to do so. These companies either have existing systems transforming into big data systems, or they are planning new applications that are big data in nature and need something 'more' than MySQL for their database platform. To grasp the advantages of distributed environment, authors in the current research aims at providing a framework for transferring data stored in Relational data storage to Decentralized structured storage system.

3 Moving from Relational Data Storage to Decentralized Structured Storage System

This section gives the details of architecture of decentralized structured storage system (Cassandra), followed by its comparison with other Relational Database Management Systems (RDBMS). It discusses moving from relational data storage (PostgreSQL database) to decentralized structured storage system (Cassandra).

3.1 Architecture of Cassandra

Cassandra is essentially a hybrid between a key-value and a column-oriented (or tabular) database. The main focus of Cassandra architecture is that the hardware failures exist and data integrity/durability should be ensured at all given times. Cassandra uses peer-to-peer distributed systems across homogenous nodes to replicate the data in all nodes in a cluster. Whenever there is a data change, each node has a sequence of commit log to execute for ensuring data consistency and all nodes have set of data at all the time. Cassandra is a row-oriented database and allows only authorized/authenticated user to connect to any node in the cluster, it uses CQL [10] (Similar to SQL) for querying data from nodes. Whenever there is a read/write request to any node, node acts as a master node for processing/executing that request and the same node decides where this request should be processed (depending on the way configuration is done for all nodes in cluster). Terminology used in Cassandra is as follows.

1. **Node:** Where the data is stored, a node can be thought of as a single physical instance where Cassandra database is mounted. It is the basic component of a distributed DBMS.
2. **Data Centers:** Data center is a collection of nodes, many nodes group together based on the data they contain and form data center
3. **Cluster:** A cluster is a group of one or more data centers.
4. **Table:** Table is a collection of ordered columns fetched by row. A row consists of columns and has a primary key. The remaining columns apart from primary key can have separate indexes and tables can be dropped, modified or inserted without impacting or interrupting updates. Example of the table formed in Cassandra is shown in Table 1.

Fig. 1 gives a snapshot of architecture of Cassandra. Architecture of Cassandra constitutes of building components, which are described as follows:

Gossip

It is a peer-to-peer communication protocol that shares location and other details of the node to other nodes in the cluster. This information is stored in each and every node in the cluster which can be used whenever any node powers out.

Table 1. Example of a table in Cassandra

Row key 1	Name	Email	Work phone	Mobile phone
	ABC	abc@pqr.com	001-123-234	912345678
Row Key 2	Name	Email	Work Phone	
	DEF	def@xyz.com	001-124-432	
Row Key 3	Name	Email		
	GHI	ghi@pqr.com		

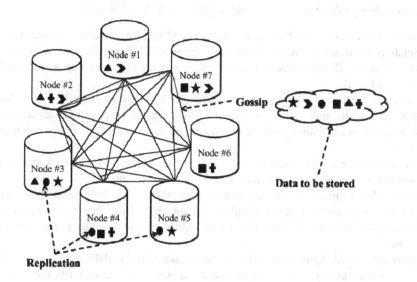

Fig. 1. Architecture of Cassandra

Partitioner
Partitioning is the 'heart' of Cassandra's architecture and used to partition the data across different nodes. Each node is responsible for processing different subset of data as assigned by partitioner. Partitioner is the hash function to compute the token of partition key. It decides which node to look at or to direct any request to (as partition key is the part of primary key in each row of data).

Replication Factor
It defines the number of replications each cluster will have. For example, replication factor of '3' means data will be replicated across nodes in the same cluster 3 times.

Replication Strategy
Replication factor determines number of copies each data will have, whereas, replication strategy determines the nodes where data has to be replicated to best suit the data availability and fault-tolerance.

Snitch

Snitch defines group of machines into data centers, which in turn is utilized by replication strategy to form the replicas and distribute them across clusters, snitch makes use of a dynamic snitch layer which choose the best replicas based on performance monitoring.

Considering Cassandra data model, data is placed in a two dimensional space within each column family. To retrieve data in a column family, users need two keys: row name and column name. In that sense, both the relational model and Cassandra are similar, although there are several crucial differences (listed below).

1. Relational columns are homogeneous across all rows in the table. A clear vertical relationship usually exists between data items; which is not the case with Cassandra columns. This is the reason Cassandra stores the column name with each data item (column).
2. In relational model 2D data space is complete. Each point in the 2D space should have at least the null value stored there. This is not the case with Cassandra, and it can have rows containing only a few items, while other rows can have millions of items.
3. In relational model the schema is predefined and cannot be changed at runtime, whereas in Cassandra users can change the schema at runtime.
4. Cassandra always stores data by sorting columns based on their names. This makes it easier to search for data through a column using slice queries. However, it is harder to search for data through a row unless we use an order-preserving partitioner.
5. Another crucial difference is that column names in RDMBS represent metadata about data, but never data. In Cassandra, the names of columns can include data. Consequently, Cassandra rows can have millions of columns, while a relational model usually has tens of columns.
6. Using a well-defined immutable schema, relational models support sophisticated queries that include JOINs, and aggregations. In relational model, users can define the schema without worrying about queries. Cassandra does not support JOINs and most SQL search methods. Therefore, schema has to be catered to the queries required by the application.

After a rigorous survey, authors presents Table 2 which compares Cassandra database, with various existing relational databases. These databases are differentiated on various parameters, such as architecture, data model, structure of queries, enterprise search, enterprise analytics, memory, security, data independence, usage and recovery. The analysis highlights the importance of Cassandra. To gain the functionalities provided by Cassandra, the current study experimented with sample data to migrate from PostgreSQL database to Cassandra.

Table 2. Comparison of Cassandra with relational database management systems

Product Capability	DataStax Enterprise Cassandra	Oracle RDBMS	Oracle MySQL	Microsoft SQL Server
Core Architecture	Masterless (no single point of failure)	Master-slave (single points of failure)	Master-slave (single points of failure)	Master-slave (single points of failure)
High Availability	Always-on continuous availability	General replication with master-slave	General read-only scale out replication; simple master-master	SQL Server replication, clustering and mirroring
Data Model	Dynamic; structured and unstructured data	Legacy RDBMS; Structured data	Legacy RDBMS; Structured data	Legacy RDBMS; Structured data
Scalability Model	Big data/Linear scale performance	Oracle RAC or Exadata	Manual sharding with MySQL	Manual sharding, general partitioning
Multi-Data Center Support	Multi-directional, multi-cloud availability	Nothing specific	Nothing specific	Nothing specific
Security	Full security support	Full security support	Full security support	Full security support
Enterprise Search	Full Solr integration	Handled via Oracle search	Full-text indexes only	Full-text indexes only
Enterprise Analytics	Integrated analytics with workload isolation with MapReduce, and Hive	Analytic functions in Oracle RDBMS via SQL MapReduce	Some analytic functions, no Hadoop support	Basic analytic functions
Database Option	Built-in in-memory option	Columnar in-memory option	MySQL cluster	Coming in-memory option
Enterprise Management & Monitoring	DataStax OpsCenter & automated management services	Oracle Enterprise Manager	MySQL Enterprise Monitor	SQL Server Enterprise Studio
Operations	No join operation, but use various other methods to show results similar to join	Performs search and insert operation (as well as various other costly operations like join)	Performs all the operations	Various operations are performed
Data Independence	Data on different data centers are independent and separable	Data depends on data types defined in the schema	Data depends on data types defined in the schema	Data depends on data types defined in the schema
Usage	Users using very large database (where not possible to store data on a SQL database)	Users of medium scale business	General users (where numbers of users are less)	General users (where numbers of users are less)
Recovery and atomicity	Remembers deletes, but full recovery is manual- using node tool	Doesn't remember delete and no chance of recovery	No provision of data recovery is present	Data can only be recovered, when log file is maintained

3.2   Moving from Relational Data Storage to Decentralized Structured Storage

Current study proposes a solution to the problem of managing growing data on various applications by migrating PostgreSQL database (centralized approach) to Cassandra (distributed approach). The whole process of migrating PostgreSQL database to Cassandra is divided in three layers as shown in Fig. 2. Firstly, authors are applying Google Snappy algorithm on the data to reduce the size of data stored in the database which directly impacts the cost of migration. After the size is reduced, an intermediate database (MySQL) is used to transfer whole database on a NoSQL database (Cassandra).

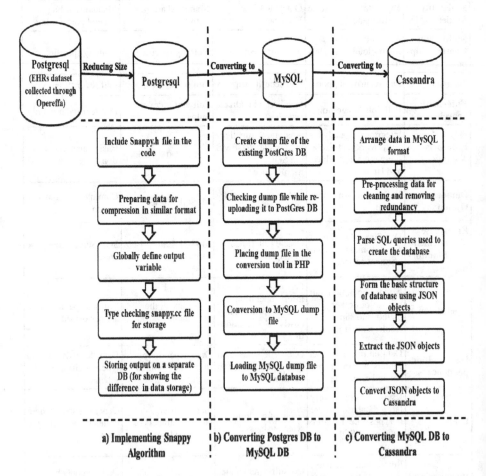

Fig. 2. Migrating from Relational data storage to Decentralized structured storage

3.3   Compressing Database

Snappy (previously known as Zippy) is a fast data compression and decompression library written in C++ by Google based on ideas from LZ77 [1, 11]. It does not aim for maximum compression, or compatibility with any other compression library; instead, it aims for very high speeds and reasonable compression. Compression speed is 250 MB/s and decompression speed is 500 MB/s using a single core of a Core i7 processor running in 64-bit mode. The compression ratio is 20–100% lower than gzip [12]. Snappy is widely used in Google projects like BigTable, MapReduce and in compression data in Google's internal RPCsystems. It can be used in open-source projects like Cassandra, Hadoop, LevelDB, RocksDB, Lucene[9]. Decompression is tested to detect any errors in the compressed stream. Snappy does not use inline assembler and is portable [13]. Fig. 3 describes the complete Google Snappy algorithm working and is delivering a final product – Snappy test file, which will take a database as an input and will compress the data to almost 80% of original data and will deliver a final product as a compressed database. For the implementation of Snappy algorithm, we need to include this 'snappyfile.h' file in our program, and it will compress the data, which is the input in the code. Snappy test file is the file used for testing of Snappy algorithm, and we need it to check, whether the compressing algorithm is working fine or not. For delivering the end product, we need to include snappy public file in the code, which defines all the functions necessary to compress the data. Apart from this, we also need 'snappy.h' file and 'snappy.cc' file, which is the header file of Snappy algorithm, and is provided open source by Google. Fig. 2(a) defines the complete process of application of Google Snappy algorithm on data.

3.4   PostgreSQL to Cassandra

The data stored in PostgreSQL database is converted into its dump files. The idea behind this dump method is to generate a text file with SQL commands. Consequently, this is fed back to the server, which will recreate the database in the same state as it was at the time of the dump. Authors used 'pg_dump' command to create the dump file of the existing database (in PostgreSQL), which we need to convert. This dump file is converted to a MySQL dump file, which can further be loaded into SQL database. This converts PostgreSQL Database to MySQL database. Fig. 2(b) describes the process of conversion of data from a PostgreSQL Database to MySQL database, considering that we should not get any loss of data during transfer.

Cassandra is a database with variable schema. It is column oriented, and gives us the flexibility to store data of different types on a same database. This gives us an advantage of storing data together. Thus, we considered moving this MySQL database to NoSQL database (Cassandra). Our work is implementation of the architecture described by Phani Krishna in [14]. To move the database from a relational model to Cassandra, authors are using ETL tools. They firstly extract the data, and transform the data by cleansing and enriching it to a processed data. Then, the data is further parsed and converted to JSON objects. Then with the help of JSON files [15], it converts the database to Cassandra. Fig. 2(c) shows the way data is moved from MySQL database to Cassandra database after parsing the queries and obtaining the JSON objects from the SQL query. Fig. 4 presents a snapshot of JSON object creator from MySQL database. These JSON objects are further converted to Cassandra and we get our desired result.

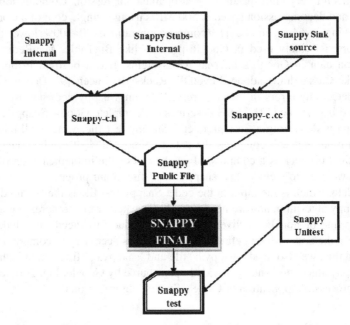

Fig. 3. File map of GOOGLE Snappy Algorithm

```
[J] SQLParserUtil.java ⊠
    package modeler;

⊕ import java.lang.reflect.Modifier;[]

    public class SQLParserUtil {

        private static final Gson GSON = new GsonBuilder().excludeFieldsWithModifiers(Modifier.TRANSIENT).create(

        public static void main(String[] args) throws StandardException {
            SQLParser parser = new SQLParser();
            StatementVisitor visitor = new StatementVisitor();
            parser.parseStatement("select t1.col2,t2.col2,t3.col3,t4.col2,t5.col1 from table1 t1, table2 t2, tabl

            System.out
        }
    }
}
```

Fig. 4. JSON objects Creator from a MySQL database

4   Experiments

For experimentation authors are using datasets of standardized Electronic Healthcare Records (EHRs) database [16-17]. Speedy access of information is highly demanded by the users now a days. There are millions of health organizations and billions of users. In such scenario, data must be stored in such a way which guarantees instant access. Availability is very critical for healthcare domain as unavailability of right information at right time may even result in loss of patient life. Other than availability; standardization, sparseness and volatility are also very critical for EHRs [18]. Several standard development organizations (openEHR, CEN, ISO and HL7) [19-23] are working to provide a standard which can be adopted by every heath organization to achieve globalization. To deal with sparseness and volatility Entity Attribute Value (EAV) model is preferred over relational model [18]. For implementing the framework proposed in previous section, we considered the database, which contains standardized EHRs data. EHRs data was synthesized by the authors using a clinical application named Opereffa [24]. Opereffa stands for openEHR REFerence Framework and Application. We explored Opereffa for our research purpose and found that it stores EHR data in a standard format (openEHR) in a single generic table (based on EAV model) using PostgreSQL. Fig. 5 provides a snapshot of the data collected through Opereffa by the authors. Database shown in Fig. 5 follows EAV model approach where "context_id" specifies the entity, "archetype path" resemble attribute column of EAV model and "value_string", "value_int" and "value_double" are analogical to the value part of EAV model. As EAV approach is followed in Opereffa, the stored dataset will be free from sparse entries. The dataset stored through Opereffa is available at a single computer where Opereffa is running. Everyone who wants to access the data needs to communicate to the single point of contact. Due to this centralized storage, availability is affected a lot. To improve on availability, we should move towards the distributed storage approach (Cassandra) by implementing the framework proposed in previous section. Sample datasets varying from 5000 to 50,000 have been used for conducting experiments for this study.

	id [PK] bigint	context_id character var	archetype_n character var	archetype_p character var	name character var	value_string character var	value_int bigint	value_double double precis	session_id character var	instance_ind integer	field_created timestamp(0	archetype_c timestamp(0	tolven_conte character var	value_at_pat character var	deleted boolean
1	19007	abc	openEHR-EHR	/data[at000	magnitude			123	0616e9c9-00	0	2014-10-28	2014-10-28			FALSE
2	19006	abc	openEHR-EHR	/data[at000	unit	/min			0616e9c9-00	0	2014-10-28	2014-10-28			FALSE
3	19005	abc	openEHR-EHR	/data[at000	value	at0006			0616e9c9-00	0	2014-10-28	2014-10-28			FALSE
4	19004	abc	openEHR-EHR	/data[at000	value	avb			0616e9c9-00	0	2014-10-28	2014-10-28			FALSE
5	19003	abc	openEHR-EHR	/data[at000	value	at0017			0616e9c9-00	0	2014-10-28	2014-10-28			FALSE
6	19002	abc	openEHR-EHR	/data[at000	value	at0012			0616e9c9-00	0	2014-10-28	2014-10-28			FALSE
7	19001	abc	openEHR-EHR	/data[at000	magnitude			123	383f858e-be	0	2014-10-28	2014-10-28			FALSE

Fig. 5. Snapshot of PostgreSQL database

5   Results

Authors have implemented Google Snappy algorithm, and have tested the data compression rate of the end product to figure out the compression rate of the algorithm.

The results obtained are shown in Table 3. To test the difference in time execution different queries have been executed by authors on standard based clinical database and the corresponding Cassandra database. To account for scalability, varying size of datasets has been considered to collect comparative results. Queries are executed considering four main parameters of any query application (insert, delete, search and select). The corresponding times taken by various queries are shown graphically in Fig. 6.

Table 3. Compression results after applying Google Snappy Algorithm

Original Data	Compressed Data
149 KB	121 KB
872 KB	536 KB
1.3 MB	807 KB

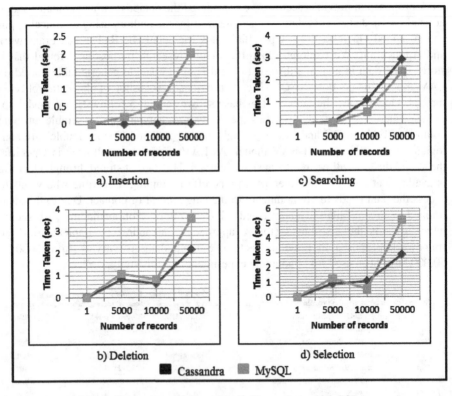

a) Insertion

c) Searching

b) Deletion

d) Selection

■ Cassandra   ■ MySQL

Fig. 6. Comparison of time taken in Cassandra and MySQL

Results obtained in Fig. 6 shows the benefits achieved in terms of time taken for executing queries on MySQL database and Cassandra. Time is calculated in MySQL, using *time()* function and in Cassandra, using *TRACE ON* command [25], which enables user to trace the amount of time an operation needs, in all its steps.

6   Conclusions and Future Scope

To account for the scalability, there is need of shifting to an approach different from relational model which can guarantee high availability with zero fault tolerance, one such model is Cassandra. Current research is a step towards providing an integrated solution for migrating data from relational data storage to decentralized structured storage system. The study implements a framework migrating from PostgreSQL to Cassandra. To minimize the cost of migration, authors in the current research implements Google snappy algorithm before migration. Table 2 is presented to give a clear differentiation of Cassandra from various existing RDBMS. An experimental comparative analysis is done to account for the benefits achieved after migrating to Cassandra from PostgreSQL in terms of time taken to execute a query (insert/ delete/ search/ select). Experiments are performed on different sizes of datasets (considering the scalability of data) ranging from 1 instance to 50,000 instances of standardized EHRs (based on openEHR standard).

In future, work can be done for providing security to all the data nodes and to provide Gossip Protocol complete information of all the nodes. A tool can also be built to change a PostgreSQL data directly to Cassandra, so that we can remove the use of MySQL database, and can get the desired result. This work is in an initial effort to move standard based clinical data to Cassandra. The researchers may apply more NoSQL databases such as MongoDB and Redis to perform a comparative analysis of all the databases. Current research can benefit other arenas related to Big Data such as meteorology, genomics, connectomics, complex physics simulations, biological and environmental research, internet search, finance and business informatics.

References

1. DataStax Corporation, White paper: Why Migrate from MySQl database to Cassandra and How? International Journal of Computer Trends and Technology 3(2) (2012)
2. Chang, F., Dean, J., Ghemawat, S., Hsieh, W.C., Wallach, D.A., Burrows, M., Chandra, T., Fikes, A., Gruber, R.E.: Bigtable: A distributed storage system for structured data. In: Proceedings of the 7th Conference on USENIX Symposium on Operating Systems Design and Implementation, vol. 7, pp. 205–218 (2006)
3. Candia, G.D., Hastorun, D., Jampani, M., Kakulapati, G., Pilchin, A., Sivasubramanian, S., Vosshall, P., Vogels, W.: Dynamo: amazonOs highly available key-value store. In: Proceedings of twenty first ACM SIGOPS symposium on Operating systems principles, pp. 205–220 (2007)
4. Lakhsman, A., Malik., P.: Cassandra - A Decentralized Structured Storage System. In: International Conference on Computing, Engineering and Information (2012)
5. Wang, G., Tang, J.: The NoSQL Principles and Basic Application of Cassandra Model. In: International Conference on Computer Science & Service System, China (2012)
6. Gilbert, S., Lynch, N.: Brewer's conjecture and the feasibility of consistent, available, partition-tolerant web services. In: vol. 33(2), pp. 51–59. Massachusetts Institute of Technology,ACM SIGACT News Homepage archiv, Cambridge (2002)
7. Theorem, B.: Practical Partition-Based Theorem Proving for Large Knowledge Bases. In: MacCartney, B., McIlraith, S., Amir, E., Uribe, T.E. (eds.) 18th Int'l Joint Conference on Artificial Intelligence, IJCAI 2003 (2003)

8. Bailis, P., Ghodsi, A.: Eventual Consistency, 'Eventual Consistency Today: Limitations, Extensions, and Beyond. ACM, UC Berkeley (2013), doi:1542-7730/13/0300

9. Featherston, D.: Cassandra: Principles and Application. In: International Conference on Computing, Engineering and Information, University of Illinois at Urbana-Champaign

10. CQL, https://cassandra.apache.org/doc/cql/CQL.html

11. PostgreSQL White Paper: How to increase performance, scalability and security within a Session Management architecture- (2005)

12. Network Defense- White Paper: Current open issues in NoSql database

13. Google Code, https://code.google.com/

14. Phani Krishna Kollapur Gandla, Migration of Relational Data structure to Cassandra (No SQL) Data structure, http://www.codeproject.com/Articles/279947/Migration-of-Relational-Data-structure-to-Cassandr

15. What is JSON, http://www.json.org

16. Beale, T., Heard, S.: The openEHR architecture: Architecture overview. In: The openEHR release 1.0.2, openEHR Foundation (2008)

17. Duftschmid, G., Wrba, T., Rinner, C.: Extraction of standardized archetyped data from Electronic Health Record Systems based on the Entity-Attribute-Value Model. International Journal of Medical Informatics 79(8), 585–597 (2010)

18. Batra, S., Sachdeva, S., Mehndiratta, P., Parashar, H.J.: Mining standardized semantic interoperable electronic healthcare records. In: Pham, T.D., Ichikawa, K., Oyama-Higa, M., Coomans, D., Jiang, X. (eds.) ACBIT 2013. CCIS, vol. 404, pp. 179–193. Springer, Heidelberg (2014)

19. OpenEHR Community (accessed 10, 2013), http://www.openehr.org/

20. CEN - European Committee for Standardization: Standards (accessed May, 09), http://www.cen.eu/CEN/Sectors/TechnicalCommitteesWorkshops/CENTechnicalCommittees/Pages/Standards.aspx?param=6232&title=CEN/TC+251

21. : ISO 13606-1.: Health informatics: Electronic health record communication. Part 1: RM, 1st edn (2008)

22. ISO 13606-2.: Health informatics: Electronic health record communication. Part 2: Archetype interchange specification, vol. 1 (2008)

23. HL7. Health level 7 (First accessed 10/13), http://www.hl7.org

24. Opereffa, http://opereffa.chime.ucl.ac.uk/introduction.jsf

25. Cassandra, Tracing on Feature, http://www.datastax.com/documentation/cql/3.0/cql/cql_reference/tracing_r.html

Comparing Infrastructure Monitoring with CloudStack Compute Services for Cloud Computing Systems

Aparna Datt[1], Anita Goel[2], and Suresh Chand Gupta[3]

[1] Department of Computer Science, University of Delhi, India
[2] Department of Computer Science, Dyal Singh College, University of Delhi, India
[3] Visiting Faculty, Department of Computer Science and Engineering, IIT Delhi, India
{aparna.datt,goel.anita,gupta.drsc}@gmail.com

Abstract. CloudStack is an open source IaaS cloud that provides compute, network and storage services to the users. Efficient management of available resources in the cloud is required in order to improve resource utilization and offer predictable performance to the customers. To facilitate providing of better quality of service, high availability and good performance; a comprehensive, reliable, centralized and accurate monitoring system is required. For this, the data needs to be collected from the components of CloudStack and analyzed in an efficient manner. In this paper, we present a detailed list of attributes required for monitoring the infrastructure associated with CloudStack. We identify the processes related with the compute services and its associated parameters that need to be monitored. We categorize the infrastructure monitoring, and list the parameters for monitoring parameters. Further, the proposed list is applied to three monitoring software that are commonly used for monitoring resources and processes associated with CloudStack. Developers and system administrators can benefit from this list while selecting the monitoring software for their system. The list is useful during the development of new monitoring software for CloudStack, as the functionality to be monitored can be selected from the list.

1   Introduction

CloudStack is an open source, open standards, multi-tenant cloud orchestration platform. It is scalable, flexible, secure and hypervisor agnostic. A service offering is a set of virtual resources, like, CPU, memory and disk. Compute and storage resources are provisioned to the users in form of service offerings. The system administrator can define different type of offerings as per the user requirements. A compute service offering includes guest CPU, guest RAM, guest networking type (virtual or direct) and tags on the root disk [14]. Along with the resources, a service offering also involves features related to resource metering, usage and charges for usage. A user can select from the available offerings while creating a virtual machine (VM). Once provisioned, CloudStack enables users with virtual machine management services that include starting, stopping, restarting and destroying virtual machines. Management of associated resources, like, CPU, network and storage and their utilization is also a part of VM management.

W. Chu et al. (Eds.): DNIS 2015, LNCS 8999, pp. 195–212, 2015.

Using virtual machines in an organization is a cost effective mechanism for management of fluctuating system workloads. With increase in demand for the compute services, size of the resources for enabling these services is also increasing. An efficient and accurate monitoring system is required to monitor resource utilization and status of the system. Monitoring of compute services basically includes monitoring the available physical infrastructure, like, CPU, memory and disk and the associated processes. The performance of infrastructure resources and allocation of these resources in form of the virtual machines needs to be measured for complete system performance.

The process of infrastructure monitoring involves collecting information about the system, and analyzing it to understand performance levels of the systems. Data required for monitoring the system resources and available services is collected at multiple points in the CloudStack software. The monitoring software collect the data, analyzes it and presents it in an understandable format, like, graphs, charts etc. for the users. Some of the popular open-source monitoring software used with CloudStack are- CollectD, Nagios and Zenoss Zenpack. These software target specific aspects of infrastructure monitoring, like, resource usage, network statistics etc. None of the available monitoring software target at the comprehensive monitoring of CloudStack.

Also, in the requirement analysis phase there is a need for designers of the system to have a detailed knowledge of the information to be monitored. While the system is running, the system administrator needs to monitor the resources and performance of the system. Instead of identifying the monitoring information at different levels by different users of the system, it would be of use if a comprehensive list of resources and processes to be monitored is available to the users. It would make it convenient to the users to select the functionality they desire to monitor from the given list.

In this paper, the aim is to identify essential features of CloudStack compute that need to be measured for infrastructure monitoring. The infrastructure functionality that needs to be monitored for CloudStack compute can be divided under four main criteria. We present a categorically distributed list of attributes relevant to compute infrastructure in CloudStack that need to be monitored for efficient functioning of the system. We have listed attributes under the following headings-

- Compute infrastructure
- Compute usage data
- OS process utilization data
- Background processes

A detailed study of CloudStack, its physical and logical components and the associated API's lead to the creation of this list [15]. This list will provide the system administrators and system developers an insight into the data that needs to be collected from the system for providing exhaustive system monitoring information. The users can use this list and make selection of required monitoring functionality. Also, third party monitoring software developers can refer to the list while updating the existing monitoring software and also during the development of new monitoring solutions for CloudStack. The attribute list presented here is a comprehensive list inclusive of all the main areas of compute monitoring in the current system and can be updated as and when new compute functionality is added to the system. Instead of generating their individual monitoring list, various users associated with the system can select the attributes from the given list.

A case study based on three popular third party monitoring tools used with CloudStack is performed. We have applied our list to CollectD [17], Nagios [18] and Zenoss Zenpack [16] to find the extent and kind of monitoring functionality provided by these infrastructure monitoring software for CloudStack.

In this paper, Section 2 provides an overview of CloudStack. Section 3 describes the structure of infrastructure monitoring along with the functionality lists. In section 4, we describe the case study for identifying functionality in existing infrastructure monitoring software. Section 5 discuses advantages of functionality list. The related work is discussed in Section 6. Section 7 concludes the paper.

2   An Overview of CloudStack

CloudStack is an open source IaaS cloud platform governed by the Apache Software Foundation. It is used for building public, private and hybrid IaaS clouds. Cloud infrastructure basically comprises of resources required for providing compute, storage and network services. On-demand, elastic cloud computing services can be provisioned using CloudStack. Service providers can sell self-service virtual machine instances, storage volumes, and networking configurations over the Internet [13]. CloudStack is a refined platform that has in-built features, like, massive scalability, high availability and centralized management, which allows single point management of vast number of physically distributed resources. CloudStack software works with a variety of hypervisors, like, VMware, Oracle VM, KVM and XenServer.

CloudStack manages its vast number of geographically distributed servers effectively. A centralized management server takes care of the infrastructure and the processes. The system is designed in a fault tolerant manner and failure of a single component does not cause the system to crash. Management server can be serviced periodically without causing any affect on the virtual machines running in the cloud. Customizable graphical user interface is available for administrators for provisioning and managing the cloud and for users for running VMs and managing VM templates. APIs are available that provide access to the management features in the GUI.

The process of VM provisioning in CloudStack is a well-defined and step-wise process. The steps for provisioning of a VM are as follows-

- Request for an instance is placed by the user.
- Operating System (Windows, Linux), Compute Offering (CPU & RAM), Disk Offering (Volume Size), Network Offering are selected.
- Instance template is copied from secondary storage to primary storage on the cluster in which the machine is to be provisioned .
- Data volume is created on primary storage for the cluster.
- An instance is created.

Once created, a VM can be started, stopped, restarted and destroyed. Service offerings of a VM can also be changed. The status of a VM for the resources it is utilizing can also be monitored within a CloudStack system.

2.1   Architectural View of CloudStack

The architectural design of CloudStack comprises of two main components, namely, the physical component which includes the cloud infrastructure and the logical component that comprises of the management Server.

The physical infrastructure of a CloudStack implementation consists of the following [14]-

- **Host:** A compute node in a cluster is called a host. The guest virtual machines created by the users are provisioned and run on a host. It is a single computer and contains resources, like, CPU, memory, storage and networking required to run a VM. With every host a hypervisor is associated for management of VMs. Multiple hosts can be interconnected via TCP/IP network.
- **Cluster:** A combination of a set of hosts and primary storage is a cluster. The hosts in a cluster are of the same kind, i.e. all hosts in a cluster have identical hardware, run the same hypervisor, are on the same subnet, and access the same shared primary storage. Virtual machine instances can be live-migrated from one host to another within the same cluster, without interrupting service to the user.
- **Pod:** A set of one or more clusters along with a switch form a pod. These are the second largest unit in the cloud infrastructure. They are invisible to users.
- **Zone:** A single or multiple pods and secondary storage combine to form a zone. Logically a zone can be compared to a single datacenter. It is the largest infrastructural unit within the CloudStack setup and aims at providing physical isolation. An end user can see the available zones in a cloud setup and has to select a zone at the time of provisioning a VM. There can be public and private zones in a cloud. Zones specific to a particular domain are private and the zones where any user can create a guest VM are public.
- **Primary storage:** It is associated with every cluster. The disk volumes for all the VMs running on hosts in a cluster are stored in the primary storage.
- **Secondary storage:** It is the storage associated with a zone. It is used for storing templates, ISO images and disk volume snapshots.

When a cloud is set up using CloudStack, storage, IP addresses and hosts are to be provisioned. For managing the resources, zones and hosts; a software component is required. Management Server is the software component of CloudStack that is used for management of the cloud resources. It has a user interface and a set of APIs that are used to configure and manage the hosted infrastructure. The management server is stateless and can be run on a server or a VM. The management server is used for managing VM provisioning, associated storage and networking on hosts. Management of snapshots, templates, ISO images and the storage is also handled by the management server. There can be a single or multiple management servers in a cloud environment. Single management server can manage multiple zones having large number of hosts. A multiple management server deployment in form of a cluster is beneficial in cases where scalability and redundancy is required.

3   Related Work

Cloud computing with its increasing complexity and vast set of resources requires monitoring as an important component. Monitoring of infrastructure and services associated with the cloud provide an in depth information on the health and performance of the system under consideration. Performance monitoring in the clouds is a focal area of research these days. Multiple monitoring software exist for infrastructure monitoring of the cloud.

There are not many monitoring systems that satisfy all the cloud administrator requirements which imposes that clouds need to be monitored by a set of monitoring systems [3].Cloud monitoring systems need to be advanced and customized to the diversity, scalability, and high dynamic cloud environment [1].

Montes et al. [10] propose a unified cloud monitoring taxonomy, based on which they define a layered cloud monitoring architecture. They implement GMonE, a general-purpose cloud monitoring tool which covers all aspects of cloud monitoring by specifically addressing the needs of modern cloud infrastructures. Meng et al. [7] propose a MaaS framework that achieves significant lower monitoring cost, higher scalability, and better multi tenancy performance. Kai et al.[5] present SCM monitoring system that collects accurate metrics from both physical and virtual resources, including the main components of Apache CloudStack (system virtual machines, Secondary Storage, Primary Storage and management servers), and makes these data easily accessible and human readable, which is quite friendly to the CloudStack users. Bellavista [2] present a novel framework for Easy Monitoring and Management of IaaS (EMMI) solutions. McGilvary et al. [6] propose Cloudlet Control and Management System (C2MS); a system for monitoring and controlling dynamic groups of physical or virtual servers within cloud infrastructures. This system allows administrators to monitor group and individual server metrics on large-scale dynamic cloud infrastructures where roles of servers may change frequently.

Monitoring solutions based on specific characteristics are also being developed. Moldovan [8], in their paper introduce MELA, a customizable framework, that enables service providers and developers to analyze cross-layered, multi-level elasticity of cloud services, from the whole cloud service to service units, based on service structure dependencies. Morariu [11] presents the design of a monitoring solution that integrates several open source tools and can assure QoS for private clouds. He et al. [4] in their paper present a novel security monitoring framework for intrusion detection in IaaS cloud infrastructures. The framework uses statistical anomaly detection techniques over data monitored both inside and outside each Virtual Machine instance. Zhang [12], presents a SLA-driven state monitoring framework for cloud service based on the matrix factorization model that contributes to improving the monitoring accuracy, and also takes less overhead during the communication.

Distributed Architecture for Resource manaGement and mOnitoring in cloudS (DARGOS), a completely distributed and highly efficient Cloud monitoring architecture to disseminate resource monitoring information has been proposed by Molina[9]. DARGOS is flexible and adaptable in nature and ensures an accurate measurement of physical and virtual resources in the cloud keeping at the same time a low overhead and enabling cloud administrators to design better cloud provisioning strategies and to avoid SLA violations.

Our study and research did not reveal any academic research study on infrastructure monitoring list for compute services in CloudStack. With the availability of this list, the designers of the system, the system administrators and the third party monitoring tool developers, all will gain insight about the features that require monitoring in CloudStack compute. The list will help them in choosing the functionality they need to monitor in the system.

4   Structure of Infrastructure Monitoring

The physical and logical infrastructure of the CloudStack setup needs to be monitored by the system administrator. Infrastructure monitoring of the system enables the system administrator to keep an updated record of the systems performance and resource utilization. This information is beneficial to the administrator as it helps to keep a check on the health and status of the system. Features, like, live migration, high availability, failure detection and scalability can be provided efficiently if complete information about total resources, resources currently under utilization and the quantity of free resources is available. The monitoring process can also enable the administrator to keep a check on resources that are running low or are underutilized. Usage trends can also be predicted by this information. System administrators can also determine information about the most preferred compute offerings by different kinds of users, size of storage users want to associate with specific offerings etc.

In this paper, we define various aspects of infrastructure monitoring keeping in mind the system administrator's perspective. We have defined the infrastructure monitoring list after studying the CloudStack architecture, infrastructure and API's [9]. We provide the infrastructure monitoring list under following four main headers:

- *Compute infrastructure list* for monitoring all attributes associated with compute offerings.
- *Compute infrastructure usage list* for assessing resource usage.
- *Background processes list* to keep track of processes running at the backend.
- *OS Process usage list* to track usage of CPU, network and memory.

4.1   Compute Infrastructure List

Compute infrastructure of CloudStack consists of the physical resources that can be virtualized. Based on the varying demands of the users, the infrastructure can auto scale. Virtual machine provisioning is the main service provided in compute offerings. Infrastructure monitoring of compute resources facilitates the system administrator to keep a track of resources being consumed and the change in workloads of the system. This information enables the system administrator in load balancing the system. Also in case, a host fails within or outside a cluster, when VM migration is required, the knowledge of current status of the system is very important. Features like high availability and scalability also require system information to be executed efficiently.

At every level of CloudStack infrastructure, there are a specific set of attributes that provide detailed information about that particular level. Attributes at Zone level are identified as id, name and description of zone. For pods and clusters also similar kind of information is maintained. The attribute list for the hosts contains host details and supported hypervisor's details. For projects and VMs, the status, resources allocated, start/ stop dates and other information is maintained.

Table 1 provides the attribute list for compute infrastructure.

Table 1. Compute infrastructure list

Levels	Parameters	Description
Zones	Id, Name, Network type, Zone token, Description, Display text	Attributes required to identify a zone.
	Zone dedicated / not dedicated	Zone is dedicated for a particular task or not.
	Domain, Domain id, Domain name	Details of domain related to a zone.
	Local storage enabled	true if local storage offering enabled, false otherwise
	Security groups enabled	true if security groups support is enabled, false otherwise
	Pod id, Pod name	Details of the pods associated with the zone
	Cluster id, Cluster name	Details of the clusters associated with the zone.
	Resources (Account, Domain, Customer, key, project) , resource id , resource type, resource details	Meta data associated with the zone (key/value pairs)
Pods	Id, Name	Attributes defining the information required to identify a Pod.
	IP(Start/ End), Gateway, Net mask	Network information with a pod.
	Zone id, Zone name	Details of the zone related to a pod.
	Cluster Id, Cluster name	Details of the cluster related to a pod.
Cluster	Id, Name, Cluster type	Details of attributes required to identify a Cluster.
	Hypervisor type	Type of hypervisor with a cluster.
	Managed state	Whether this cluster is managed by CloudStack.
	Memory over commit ratio	Memory over commit ratio of cluster
	Pod id, Pod name	Pod related to a cluster.
	Zone id, Zone name	Zone related to a cluster.
Domain	Id, name, level	Details of attributes required to identify a domain.

Table 1. (*continued*)

	Has child	Whether the domain has one or more sub-domains
	Network domain	Network information with a domain.
	Parent domain (id, name)	Details of the parent domain, in case the domain under consideration is a child domain.
	Path	Path of the domain
Host	Id, name, State (connected/ disconnected), Removed, HA, type, version, created, capabilities	Attributes defining the information required to identify a host.
	Cluster (Id, name, type)	Cluster information related to a host.
	Hypervisor, Hypervisor version	The host hypervisor
	Management server id	Management server ID of the host
	Network kbs read, Network kbs write	Network information of host.
	Os category id, Os category name	The OS category ID of host, the OS category name of host
	Pod id, Pod name	Pod related to the host.
	Zone id, zone name	Details of zone related to the host.
	Job id, Job status	ID and status of latest asynchronous job acting on this object.
Account	Account (Id, details, type, Name, State)	Attributes defining the information required to identify an account.
	Default zone id	Default zone of the account
	Domain(name, id)	Domain related to account.
	Groups	List of groups that account belongs to.
	Project (available, limit, total)	Projects associated with an account.
	Received bytes, Sent bytes	Network traffic information of the account.
	Snapshot (available/ limit/ total)	The total number of snapshots available, can be stored and are currently stored for this account. stored by this account
	Template (available/ limit/ total)	The total number of templates available to be created, available for creation, have been created by this account.
	Vm (available, limit, total, running, stopped)	The total number of virtual machines available, can be deployed, currently running and stopped for this account.
	user(id, account, account id, account type, created, domain, domain id, user details)	List of users associated with account.
Project	Id, Display text, Name, State	Attributes defining the information required to identify a project.
	Account	Account name of the project's owner
	Domain (name, id)	Domain information of a project.

Table 1. (*continued*)

	Snapshot (available, limit, total)	The total number of snapshots available, stored and can be stored for this project.
	Template (available, limit, total)	Total number of templates available to be created, can be created and have been created by this project.
	VM (available, limit, running, stopped, total)	Total number of virtual machines available, can be deployed, currently running and stopped for this account.
	Resource (id, type)	Id and type of resource.
VM	Id, Account, Created, Details, display name, Display vm, state	Details of attributes required to identify a virtual machine.
	Domain, Domain id	The domain information with a VM.
	Group (name, id)	The group information with a VM.
	Guest os id	Os type ID of the virtual machine
	High availability enable, scalable	True if high-availability is enabled, false otherwise
	Hosted, Hostname	Host information with a VM.
	Hypervisor	Hypervisor on which template runs
	Memory, root volume	Memory allocated for a VM
	Project (name, id)	Project information with a VM.
	Service offering (id, name)	Details of service offering with a VM.
	Resource id, resource type	Resources associated with a VM.
	Template display text	Alternate display text of the template for VM
	Template (id, name)	Template information with a VM.
	Zone (id, name)	Zone information with a VM.

4.2   Compute Infrastructure Usage List

The information regarding the usage of compute infrastructure requires consistent monitoring. The current state of zones, pods, clusters, status of resources (CPU, memory and network) currently in use on hosts, number of VMs on the host etc. need to be monitored. Monitoring of the data usage helps to identify the total resource usage, quantity of available resources and current load on the system at all different levels. This information provides a detail on the usage aspect of the system.

Compute usage list contains attributes required by the administrator for monitoring the usage of the system. Information, like, host uptime, number of network read/writes, available services on each level, list of compute offerings etc. is maintained. This information provides the administrator with exact state of a system at a given point of time. Administrator can assess the utilization levels of the system, trending resources and need for more resources in the system. Table 2 displays the list for compute usage attributes.

Table 2. Compute infrastructure usage list

Level	Parameter	Description
Zones	Capacity (total, used, percent)	Total, in use and used percentage capacity of zone.
	Allocation state	Allocation state of the cluster.
Pods	Capacity (total, used, percent, type)	Total, in use, used percentage and type of capacity of Pod.
	Allocation state	Allocation state of Pod.
Cluster	Allocation state	Allocation state of cluster.
	CPU over commit ratio	CPU over commit ratio of cluster.
	Capacity (total, used, percent, type)	Total, in use, used percentage and type of capacity of the Pod.
Host	Average load	Average CPU load on the host.
	Events	Events available for host.
	Enough capacity	True if this host has enough CPU and RAM capacity to migrate a VM to it, false otherwise.
	local storage active	True if local storage is active, false otherwise.
	Last pinged	The date and time the host was last pinged.
	Resource state	Resource state of the host.
	Suitable for migration	True if this host is suitable (has enough capacity and satisfies all conditions) to migrate a VM to it, false otherwise.
Account	IP(available, limit, total)	The network information associated with the account.
	Primary storage (available, limit, total)	Total primary storage space (in GiB) available to be used, can be owned and is owned for this account.
	Secondary storage (available, limit, total)	Total secondary storage space (in GiB) available, can be owned and is owned for this account.
	Volume (available, limit, total)	Total volume available, being used and can be used for this account.
Project	Primary storage (available, limit, total)	Total primary storage space (in GiB) available to be used, can be owned, and owned for this project.
	Secondary storage (available, limit, total)	Total secondary storage space (in GiB) available to be used, can be owned, and owned for this project.
	Volume (available, limit, total)	Total volume available for this project, can be used and currently being used by this project.

4.3 OS Process Utilization Data List

Within each level of CloudStack compute, there are three standard OS processes that run to facilitate the smooth functioning of compute services. OS processes utilization data list provides the parameters that need to be monitored for measuring utilization for each of these. Utilization of CPU, memory and network is measured at host, account, project and VM level. At all levels, CPU, network and memory utilization data can be monitored for total, allocated and used options.
Table 3 presents the OS processes utilization list.

Table 3. OS process utilization list

Level	Process	Details
Host	CPU utilization	CPU (allocated, number, sockets, speed, used)
	Memory utilization	Memory (allocated, total, used)
	Network utilization	Network kbs (read, write)
Account	CPU utilization	CPU (available, limit, total)
	Memory utilization	Memory (available, limit, total)
	Network utilization	Network (available, domain, limit, total)
Project	CPU utilization	CPU (available, limit, total)
	Memory utilization	Memory (available, limit, total)
	Network utilization	Network (available, limit, total)
VM	CPU utilization	CPU(number, speed, used)
	Memory utilization	Disk – I/O (read/write), kbs (read/write), offering (id, name)
	Network utilization	Network kbs(read, write)

4.4 Background Process Functionality List

For listing the functionality and the parameters to be monitored for background processes in CloudStack compute, we have defined the background processes functionality list. This list can be referred by the system administrator for monitoring background processes and can also be utilized at the time of developing monitoring component for CloudStack.

Background processes are carried out by the management server, console proxy VM and secondary storage VM. Information required for monitoring running processes, like, resource provisioning, snapshot management, template management, zone management, orchestration, high availability, live migration, connection management and secondary storage management can be collected from the system. Parameters to be monitored within each process are identified and listed.

Management of provisioning of resources, snapshot, zone and template management are the basic processes that run in the management server to facilitate the IaaS services provided by CloudStack. Orchestration is an important process that runs in the management server. High availability and live migration are two processes that make CloudStack services more efficient and reliable.

Connection management of a VM to display VMs console on the management server web interface for remote management is managed by console proxy VM. Secondary storage VM runs secondary storage management process to facilitate snapshots, templates, volumes and ISO files management in and out of the cloud environment in a secure manner. Table 4 provides a list of background processes that need to be monitored.

Table 4. Background process list

Process	Details	Component
Resource provisioning	VM allocation to host, storage devices, IP addresses.	Management Server
Snapshot management	Full snapshots, incremental snapshots.	Management Server
Template management	Volume templates.	Management Server
Zone management	Management of zones.	Management Server
Orchestration	VM starts, snapshot copies, template propagation.	Management Server
High Availability	VMs, hosts, primary storage, secondary storage.	Management Server
Live Migration	Migration of a running VM. Stopped VM's can't be migrated.	Management Server
Connection management	Facilitates the secure connection to individual VM's through their respective hypervisor hosts to display the VMs console on the management server web interface for remote management.	Console Proxy VM
Secondary storage management	Facilitates moving snapshots, templates, volumes and ISO files in and out of the cloud environment in a secure manner. Connects to individual hypervisors to instruct them on how to mount secondary storage as well as how to store and cleanup snapshots and templates.	Secondary storage VM

5   Case Study

As a case study, we have applied these lists on three popular open source monitoring software generally used with CloudStack, namely, Zenoss Zenpack, CollectD and Nagios. This case study will enable us to identifying the kind of infrastructure monitoring provided by these tools for CloudStack. Comparison of our lists with the information monitored by these software will also help to identify the extent of monitoring depth provided by the tools. Also the case study can be used to identify the areas of CloudStack that are not being monitored effectively.

Zenoss ZenPack [16] is an open source cloud monitoring software that provides monitoring support for CloudStack. Zenoss allows system administrators to monitor some of the important aspects of CloudStack like availability, inventory, configuration, performance, and events related to the system. The key metrics provided by zenoss are for CPU and memory tracking. These metrics are provided within Zenoss at all the levels of granularity. Alerts regarding low and high thresholds of the resources are also available.

CollectD [17] is a daemon which collects information about zones, pods, clusters, storage and hosts. Information about the available and resources being used currently by the system are collected by CollectD. System performance statistics are also gathered periodically and can be stored in a variety of ways. Information to be monitored is collected in form of system metrics relating to memory, CPU, secondary storage etc. and can be used to measure the current status of system.

Nagios[18] is a popular open source cloud monitoring software. It monitors usage of available resources like CPU, memory, etc. Cloud capacity including memory, storage, private and public ip's is also monitored. Information regarding state and status of VM's is monitored. Checks for usage of various resources like memory, CPU, network and disk usage within a VM are also carried out.

Table 5 presents the list detailing comparison of monitoring of background processes by the three monitoring software under consideration. Table 6 combines the list of comparison of three selected monitoring software for compute infrastructure attributes and compute usage. OS processes Utilization data of infrastructure monitoring is displayed in Table 7. In the tables, notation '√' denotes that parameter is supported by the monitoring software; 'x' denotes parameter not supported.

Table 5. Background process case study

Process	Zenoss	CollectD	Nagios
Resource provisioning	√	√	√
Snapshot management	×	×	×
Template management	×	×	×
Zone management	×	×	×
Orchestration	×	×	×
High Availability	×	×	×
Live Migration	×	×	×
Connection management	×	×	×
Secondary storage management	×	×	×

Table 6. Compute infrastructure and usage case study list

Levels	Infrastructure Parameters	Usage Parameter	Zen oss	Zen oss	Col lect D	Col lect D	Nag ios	Nag ios
Zones	Details	Capacity (total, used)	×	×	√	√	√	√
	Resources, resource id, resource type, resource details	Percentage used	×	×	√	√	√	√
	List of Domains	Allocation state	×	×	×	×	×	×
	List of Pods		×	×	×	×	×	×
	List of Clusters		×	×	×	×	×	×
Pods	Details	Capacity (total, used)	×	×	√	√	√	√
	Network Details	Percentage used	×	×	√	×	√	√
	Zone Details	Allocation state	×	×	√	×	×	×
	List of Clusters		×	×	√	×	×	×
Cluster	Details	Type	×	×	√	×	×	×
	Hypervisor type	Capacity (total, used)	×	×	×	√	×	√
	Managed state	Percentage used	×	×	×	×	×	×
	Pod details	Allocation state	×	×	√	×	×	×
	Zone details		×	×	√	×	×	×
Domain	Details		×	×	√	×	√	×
Host	Id, name, State (connected/disconnected), Removed, HA, type, version, created, capabilities	Average load	×	×	√	×	√	×
	Cluster (Id, name, type)	CPU(allocated, number, used, speed)	×	×	√	√	√	√
	Hypervisor, Hypervisor version	Capacity	×	×	×	√	×	×
	Management server id	Last pinged	√	×	√	×	×	×

Table 6. (*continued*)

Category	Item	Attribute						
Account	Network details	Resource state	×	×	√	√	√	√
	Pod details	Suitable for migration	×	×	√	√	√	×
	Zone details		×	×	√	×	×	×
	Account details	IP(available, limit, total)	×	×	√	√	×	√
	Domain details	Memory (available, limit, total)	×	×	√	√	×	√
	Network details	Network (available, domain, limit, total)	×	×	×	√	×	√
	Project details	Primary storage (available, limit, total)	×	×	×	√	×	√
	Snapshot details	Secondary storage (available, limit, total)	×	×	×	√	×	√
	Template information	Volume (available, limit, total)	×	×	×	√	×	√
	VM (available, limit, total, running, stopped)		×	×	×	√		
	User information		×	×	×	×	×	×
Project	Id, Display text, Name, State	CPU (available, limit, total)	×	×	√	√	√	√
	Domain (name, id)	Memory (available, limit, total)	√	√	√	√	√	√
		Network (available, limit, total)	×	√	×	√	×	√
	Snapshot (available, limit, total)	Network (available, limit, total)						

Table 6. (*continued*)

	Vm (available, limit, running, stopped, total)	Secondary storage (available, limit, total)	√	√	×	√	×	√
	Resource (id, type)	Volume (available, limit, total)	×	√	×	√	×	√
		Vpc (available, limit, total)	×	√	×		×	√
VM	Details	Status	√	√	√	√	√	√
	Domain, Domain id	Memory, root volume	√	√	×	√	×	×
		Network (kbsread, kbswrite)	×	√	×	√	×	×
	Guest os id							
	High availabilty enable, scalable	Resource usage	×	√	√	√	×	×
	Hosted, Hostname		√	×	×	×	×	×
	Hypervisor			×	×	×	×	×
	Project (name, id)		√	×	×	×	×	×
	Service offering (id, name)		√	×	√	×	√	×
	Resource id, resource type		√	×	×	×	√	×
	Template (id, name)		×	×	×	×	×	×
	Zone (id, name)		√	×	×	×	×	×

The case study carried out above indicates that the major aspect of monitoring covered by all the three monitoring software is of compute infrastructure and its usage. Except for resource provisioning, monitoring support for other background processes is not there in any of the software. Our study reveals that the maximum monitoring support for OS processes is provided by Zenoss Zenpack. The common parameters monitored by all the monitoring software are the status of total resources, resources currently in use, available resources and metrics about resources at different levels of granularity.

Table 7. OS processes case study list

Level	Process	Zenoss	CollectD	Nagios
Host	Cpu utilization	√	√	×
	Memory utilization	√	√	×
	Network utilization	√	√	×
Account	Cpu utilization	√	×	×
	Memory utilization	√	×	×
	Network utilization	√	×	×
Project	Cpu utilization	√	×	×
	Memory utilization	√	×	×
	Network utilization	√	×	×
VM	Cpu utilization	√	√	√
	Memory utilization	√	√	√
	Network utilization	√	√	√

6   Benefits of Functionality List

The functionality list presented in our paper is exhaustive for listing monitoring functionality in CloudStack. We have arrived at the functionality list after performing a study of CloudStack architecture and API's. The functionality list is beneficial to the system administrators and the designers of the system. The list makes the task of monitoring easier for the system administrator as the desired features to be monitored can be selected form the defined list. Monitoring tools specific to monitoring of CloudStack compute services can also be developed by referring to this list Additions to the existing monitoring software can also be made to increase their monitoring support towards CloudStack compute. On the basis of case study presented above, it is evident that none of the existing monitoring tools available provide a complete monitoring coverage of all features of CloudStack compute. All the basic set of infrastructure, associated parameters and the processes that need to be monitored consistently for maintaining compute system efficiently are provided in this functionality list.

7   Conclusion

In our paper, we have presented a list defining the attributes and processes and their parameters of infrastructure that need to be monitored in a CloudStack compute system. It facilitates designers in specifying monitoring requirements during requirement specification phase. Also, the system administrator can use the list for selecting the functionality that needs to be monitored in the software. user with a list of all attributes that require monitoring. The process of selecting from the functionality list is easier than creation of a new list by different users of the system who require monitoring functionality. Software designers, system administrators and developers of monitoring systems for CloudStack can view and select the details of all the functionality, processes and attributes that can be monitored and features that can be integrated in any new monitoring software being developed. This list can be extended and updated when new compute services are added to the system or the current ones are updated.

References

1. Alhamazani, K., Ranjan, R., Mitra, K., Rabhi, F., Khan, S., Guabtni, A., Bhatnagar, V.: An Overview of the Commercial Cloud Monitoring Tools: Research Dimensions, Design Issues, and State-of-the-Art, http://www.nist.gov/itl/cloud/
2. Bellavista, P., Giannelli, C., Mattetti, M.: A practical approach to easily monitoring and managing IaaS environments. In: IEEE Symposium on Computers and Communications (ISCC) Symposium (2013)
3. Carvahlo, M., Esteves, R., Rodrigues, G., Granville, L.: A cloud monitoring framework for self-configured monitoring slices based on multiple tools. In: 9th International Conference on Network and Service Management (CNSM), Zurich, pp. 180–184 (2013)
4. He, S., Ghanem, M., Guo, L., Guo, Y.: Cloud Resource Monitoring for Intrusion Detection. In: IEEE International Conference on Cloud Computing Technology and Science (2013)
5. Kai, L., Weiqin, T., Liping, Z., Chao, H.: SCM: A Design and Implementation of Monitoring System for CloudStack. School of Computer Engineering and Science, Shanghai University. IEEE (2013)
6. McGilvary, A., Rius, J., Goiri, N., Solsona, F., Barker, A., Atkinson, M.: C2MS: Dynamic Monitoring and Management of Cloud Infrastructures. In: IEEE International Conference on Cloud Computing Technology and Science (2013)
7. Meng, S., Liu, L.: Enhanced Monitoring-as-a-Service for Effective Cloud Management. IEEE Transactions on Computers 62(9) (September 2013)
8. Moldovan, D., Copil, G., Truong, H., Dustdar, S.: MELA: Monitoring and Analyzing Elasticity of Cloud Services. In: IEEE International Conference on Cloud Computing Technology and Science (2013)
9. Molina, J., Vega, J., Soler, J., Corradi, A., Foschini, L.: DARGOS: A highly adaptable and scalable monitoring architecture for multi-tenant Clouds. Future Generation Computer Systems 29, 2041–2056 (2013)
10. Montesa, J., Sánchez, A., Memishic, B., Pérez, M., Antoniu, G.: GMonE: A complete approach to cloud monitoring. Future Generation Computer SystemsFuture Generation Computer Systems 29, 2026–2040 (2013)
11. Morariu, O., Borangiu, T., Morariu, C.: Multi Layer QoS Monitoring in Private Clouds. In: 24th International Workshop on Database and Expert Systems Applications (2013)
12. Zhang, Y., Liu, H., Lu, Y., Deng, B.: SLA-Driven State Monitoring for Cloud Services. In: IEEE International Conference on High Performance Computing and Communications & 2013 IEEE International Conference on Embedded and Ubiquitous Computing (2013)
13. http://cloudstack.apache.org/
14. http://docs.cloudstack.apache.org/projects/cloudstack-administration/en/4.4/
15. http://cloudstack.apache.org/docs/api/
16. http://wiki.zenoss.org/ZenPack:CloudStack
17. http://collectd.org/
18. http://exchange.nagios.org/directory/Plugins/Cloud/nagios-2Dcloudstack/details

Efficiency of NoSQL Databases under a Moderate Application Load

Mohammad Shamsul Arefin[1], Khondoker Nazibul Hossain[1],
and Yasuhiko Morimoto[2]

[1] Computer Science and Engineering Department,
Chittagong University of Engineering and Technology,
Chittagong-4349, Bangladesh
{sarefin_406,nazibul_14}@yahoo.com
[2] Graduate School of Engineering, Hiroshima University, Japan
morimoto@mis.hiroshima-u.ac.jp

Abstract. Consider the fact, that the concepts of NoSQL databases have been developed and recently, big Internet companies such as Google, Amazon, Yahoo!, and Facebook are using NoSQL databases. Although the primary focus of NoSQL databases is to deal with huge volume of heterogeneous data, these can also be suited for handling moderate volume of data, especially if the data are heterogeneous and there are frequent changes in data. Considering this we consider the development and implementation of an application with moderate volume of heterogeneous data using a NoSQL database. We perform comparative performance analysis with a relational database system. The experimental evaluations show that NoSQL databases are also often suitable for handling moderate volume of data.

Keywords: NoSQL database, relational database, data distribution, mobile server.

1 Introduction

To overcome the limitations of RDMS in handling large volume of data, NoSQL systems were developed. Although NoSQL databases were developed to deal with huge volume of data, they can also be efficient for the systems those deal with moderate size of heterogeneous data. In addition, NoSQL databases can be efficient for developing secure applications and applications those deal with frequent changes of data.

In this paper, we present an application framework for the entrance examinations at universities in Bangladesh using a NoSQL database and show that NoSQL database systems are suitable and are comparable to relational database systems for handling some applications.

The main contributions of this paper can be summarized as follows: At first, we design the database architecture of the system using a NoSQL database. Then, we develop a platform independent interface so that the users are able to interact with the system using any type of devices such as PC, smart phones,

W. Chu et al. (Eds.): DNIS 2015, LNCS 8999, pp. 213–227, 2015.

feature phones, PDA etc. Finally, we perform several experiments to show the effectiveness of our system over relational database systems.

The remainder of this paper is organized as follows. Section 2 provides a brief review of related works. In section 3, we detail the framework of our developed system. Section 4 presents the experimental results. Finally, we conclude and sketch future research directions in Section 5.

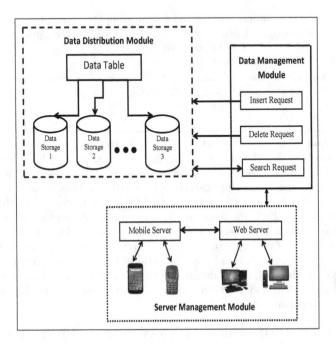

Fig. 1. System architecture of the application

2   Related Works

Although many companies continued their research about efficient handling large volume of data, Google is the pioneer in this regard by publishing the Bigtable paper [1]. Later on Amazon published Dynamo [2] that described several technologies all of which together have come to be known as NoSQL. Based on the concepts of NoSQL, several database systems such as MongoDB [3], Cassandra [4], HBase [5], Neo4j [6] were developed.

Later researchers performed different types of analysis on NoSQL databases. Harter et al. [7] presents an analysis on HBase and HDFS considering the messages of Facebook. The goal of this analysis was to find the effectiveness of HDFS storage in HBase. In this analysis, they described multilayer storage techniques. They argue that writing is expensive in disk level due to logging, compaction,

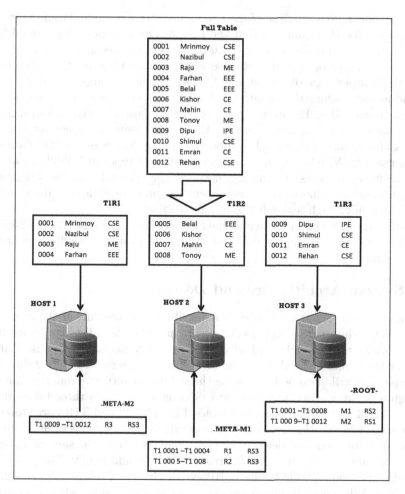

Fig. 2. Splitting data into multiple servers

replication, and caching. Details about HBase system is presented by Dimiduk et al. [8]. In this paper the authors briefly described the architecture, data distribution techniques, map reduce function. Kaur et al. [9] worked on distributed database system and described it with the concurrency control algorithms. The also discussed distributed 2PL, wound-wait, basic timestamp ordering and distributed optimistic algorithms.

Stevic et al. [10] analysed performance on storing documents which is unstructured. They argue that RDBMS is not efficient enough for storing unstructured documents. They also suggest to use NoSQL databases for handling large amount of documents. Franke et al. [11] explained the techniques of semantic web data management using HBase and MySQL Cluster. They designed a novel database schema for HBase and generic RDF database schema for MySQL Cluster. After the experiment, a comparison of the two approaches was made. From the experimental results, they conclude that the HBase is capable of dealing with larger RDF datasets and are superior in both query performance and scalability.

Different types of NoSQL database systems with their relative advantages and disadvantages are discussed in [12]. Here, the author argued that if data is bound in one machine then relational database systems are sufficient to manage data with scalability. However, the use of NoSQL database systems can increase the performance significantly. To fulfil the requirements of present stage a new trends of database is discussed by Pokorny [13]. A survey on NoSQL database is discussed by Vogels [14] with describing different types of NoSQL databases with their key features and implementation. The survey found the strength of NoSQL databases due to their easy operation procedure, flexible data model, high availability, high scalability and fault tolerance.

In this paper, we provide the implementation of an application using a NoSQL database and show the applicability of NoSQL databases in handling moderate data volume.

3   System Architecture and Design

The architecture of our application is a distributed architecture as shown in Figure 1. The architecture comprises three main modules: data distribution module, data management module, and server management module. Data distribution module is responsible for distributing data among the servers. Figure 2 shows an example of distribution of data among three different servers. From the example of Figure 2, we can see that the data table is divided into three tables name T1R1, T1R2 and T1R3. Then, the tables T1R1, T1R2 and T1R3 are stored in three servers Host1, Host2 and Host3, respectively. Data management module is responsible for insertion, deletion and search of data from the servers. In order to insert data, at first, we need to create table and add family. The procedure for table creation and add family in HBase are given in Table 1 and Table 2, respectively. When data insertion request is made, a key-value instance is created and put it to a database table using table name, row key, column family and column qualifier. Table 3 shows the insertion procedure of values in a table.

If search request is made, data is retrieved with the help of algorithms as shown in Table 4 and Table 5. The algorithm of Table 4 is used for retrieving a row while the algorithm in Table 5 is used to retrieve values. Figure 3 shows an example of searching process. When it is necessary to search a particular information say "row 0009" by a a client, the client first asks the server for the location of -ROOT- table. Server the sends location information of the -ROOT- file. Then, -ROOT- table is accessed for the name and location of .META file that contains the information about the particular row. The .META file is then accessed for the location of the particular row. From .META file, we get the table name and table server location. Then, the table of the specified server is accessed and read the particular row.

The algorithm for deleting a value from a table is given in Table 6, whereas the algorithm for deleting a row from a table is given in Table 7.

Table 1. Algorithm for creating a table

```
Algorithm_CreateTable(tablename){
    1. create HBaseConfiguration object 'hc'
    2. create HBaseAdmin object 'hba' with parameter 'hc'
    3. create HTableDescriptor object 'ht' with parameter 'tablename'
    4. if table not exist, then call createTable() method of 'ht'
         to create table
    5. return
}
```

Table 2. Algorithm for adding family in a table

```
Algorithm_AddFamily(tablename, familyname){
    1. create HBaseConfiguration object 'hc'
    2. create HBaseAdmin object 'hba' with parameter 'hc'
    3. create HTableDescriptor object 'ht' with parameter 'tablename'
    4. if table exist then
        4.1 call  hba.disableTable(tablename) method to disable table
        4.2 create HColumnDescriptor object 'cf' with parameter familyname
        4.3 add family by method  hba.addColumn(tablename, cf)
        4.4 enable table by method hba.enableTable(tablename)
    5. return
}
```

Table 3. Algorithm for adding values in a table

```
Algorithm_AddValue(tablename,rowkey,clmfamily,clmquantifier,value){
    1. create HTablePool object 'pool'
    2. create HTableInterface object 'userTable' by calling method
         pool.getTable(tablename)
    3. create Put object 'p' with parameter 'rowkey'
    4. add value in the Put object by method
         p.add(clmfamily, clmquantifier, value)
    5. add 'p' in 'userTable' by method usersTable.put(p)
    6. return
}
```

Table 4. Algorithm for reading a row from a table

```
Algorithm_ReadRow(tablename, rowkey){
   1. create HTablePool object 'pool'
   2. create HTableInterface object 'userTable' by calling method
        pool.getTable(tablename)
   3. create Get object 'g' with parameter 'rowkey'
   4. create Result object 'r' by calling method  usersTable.get(g)
   5. for KeyValue object 'kv' to  r.raw() do
     5.1 get the row by method kv.getRow()
     5.2 get the family by method kv.getFamily()
     5.3 get the qualifier by method kv.getQualifier()
     5.4 get the value by method kv.getValue()
   6. close 'usertable' by method usersTable.close()
   7. return
}
```

Table 5. Algorithm for reading a value from a table

```
Algorithm_ReadValue(tablename, rowkey,clmfamily,clmquantifier){
   1. create HTablePool object 'pool'
   2. create HTableInterface object 'userTable' by calling method
        pool.getTable(tablename)
   3. create Get object 'g' with parameter 'rowkey'
   4. create Result object 'r' by calling method  usersTable.get(g)
   5. get the value by method r.getValue(clmfamily,clmquantifier)
   6. return
}
```

Table 6. Algorithm for deleting a value from a table

```
Algorithm_DeleteValue(tablename,rowkey,clmfamily,clmquantifier){
   1. create HTablePool object 'pool'
   2. create HTableInterface object 'userTable' by calling method
        pool.getTable(tablename)
   3. create Delete object 'd' with parameter 'rowkey'
   4. add this in the Delete object by method
        d.deleteColumns(clmfamily, clmquantifier)
   5. delete value by the method usersTable.delete(d);
   6. close 'usertable' by method usersTable.close()
   7. return
}
```

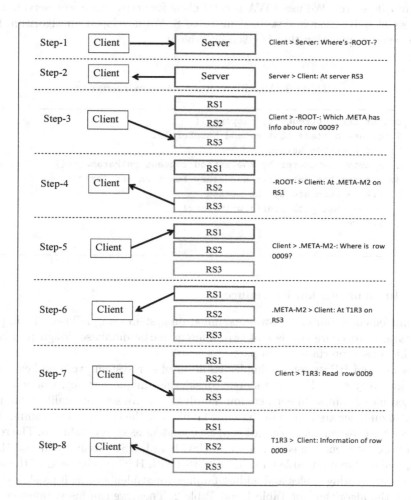

Fig. 3. Search procedure of our system

Table 7. Algorithm for deleting a row from a table

```
Algorithm_DeleteRow(tablename,rowkey){
    1. create HTablePool object 'pool'
    2. create HTableInterface object 'userTable' by calling method
       pool.getTable(tablename)
    3. create Delete object 'd' with parameter 'rowkey'
    4. delete value by the method usersTable.delete(d);
    5. close 'usertable' by method usersTable.close()
    6. return
}
```

The server management module deals with two different servers: web server and mobile server. We use JAVA *servlet* class for activating a web server. The process of web server is described in Table 8. We developed an algorithm for SMS server application that is given in Table 9.

Table 8. Algorithm for activating web server

```
Algorithm_servlet(request, response){
    1. create a class that extend HttpServlet
    2. add Override method
    3. receive parameter by the method request.getParameter()
    4. Process the data with necessary function and logic
    5. create response by the method
        response.getWriter().write(msg)
    6. return
}
```

3.1   Implementation Procedure

We implemented our cluster-wise examination system using HBase. The implementation procedure consists of two different parts: database design part and graphical user interface design part.

As we used HBase for the implementation of our system, we used key-value pair for every data. Data are stored considering unique row key. Columns are divided into families. In each column family, there are several qualifier. Similar types of qualifier are placed in the same family. For the cluster-wise examination system, we consider the information of each student as shown Table 10. The row-key of each student is a combination of HSC board, HSC passing year and HSC roll number. We used table name that relates with HSC passing year of the students. For creating tables and adding families for students of each passing year, we call the algorithms of Table 1 and Table 2. Then, we can insert information with the help of the algorithm of Table 3. Table 11 shows the information that is considered for each university. Here, we consider university login name as the the row key. For the departments within the university, we store the information as shown in Table 12. For each university, we consider a department table. So, the number of department tables is equal to the number of universities.

Table 13 listed all the examination categories information. Here, the row-key is the category name. The list of departments those are within same category are recorded in this table. Names are separated by commas. For each category, a new table will be created where all the necessary information of that category is recorded. The row key for this table is students row key as this table lists the information of the students based on their application category.

Table 9. Algorithm for SMS server application

```
Algorithm_SMSmanage(){
  1. create a class that extend BroadcastReceiver
  2. Override the method
       onReceive(Context context, Intent intent)
  3. if intent.getAction() ==
       android.provider.Telephony.SMS_RECEIVED then
    3.1 create String object 'message'
    3.2 create a Bundle object 'bundle' by the method
           intent.getExtras()
    3.3 if bundle is not null then
        3.3.1 create Object array 'pdus' by the method
                  bundle.get("pdus")
        3.3.2 create SmsMessage array 'chunks' by passing
                  parameter pdus.length
        3.3.3 for 0 to pdus.length-1 {
                  chunks[i] = SmsMessage.createFromPdu((byte[]) pdus[i]);
                  number = chunks[i].getOriginatingAddress();
                  message += chunks[i].getMessageBody();
                  }
        3.3.4 send message to server
        3.3.5 receive response string
        3.3.6 create SmsManager object 'manager'
        3.3.7 send message by the method
              manager.sendTextMessage(number, null,
                            reply message, null, null);
        3.4 end if
  4. end if
  5. return
}
```

The admission choice list of each students are also recorded in the category list. This information is necessary to determine the department for the students based on the examination results. We also consider a table for keeping the records of administration as listed in Table 14.

In our developed system, there are the facilities for students login, university login, and administrator login. Students can see the options for admission, payment procedure, result, choice list after login in the system. Here, they can modify their choice list until a certain period of time. They can also appear at different types of MCQ and written practice tests. The system also has the facility of automated score generation for both MCQ and written tests. For generating scores of MCQ type questions, we just compare users answers with the answers of the database. However, for computing the scores of written type questions, we use string matching algorithm that compares the answer of a question stored in the database with the answer of the users. The system also has the

Table 10. Students' information

Table Name	Column Family	Column Qualifier
Student Information	Personal Information	Student Name
		Father Name
		Mother Name
		Gender
		Birthdate
		Mobile No.
	HSC Information	Board
		Roll
		Registration
		Group
		Session
		Type
		GPA
		GPA without Optional
		Subject 1
		Subject 2
		Subject 3
		Subject 4
		Subject 5
		Subject 6
		Subject Optional
	SSC Information	Board
		Roll
		Registration
		Group
		Session
		Type
		GPA
		GPA without Optional
	Applicable Category	Category 1
		Category 2
		•
		•
		•
		Category N

facility to make the students aware about any post or update in the university information.

After login in the system by an university representative, he is able to post or update required information in the system. Administrator of the system has all types of rights over the system.

Our developed system is also accessible by the students via text message. For accessing our system using text message, a student needs to enter the information in the format "[BOARD NAME][SPACE][PASSING YEAR][SPACE][ROLL NO]

Table 11. Universities information

Table Name	Column Family	Column Qualifier
Universities	General Information	University Name
		Login Name
		Password
		University Address
		Faculty Number
		Department Number

Table 12. Department information

Table Name	Column Family	Column Qualifier
Each University Information	General Information	Department Name
		Faculty Name
		Category Name
		Total Seat
	Requirements	Admission test Result
		HSC result
		Other Requirments

Table 13. Category information

Table Name	Column Family	Column Qualifier
Each Category Name	General Information	Application No
		Payment
		Exam Roll
		Exam Seat
	Result	Total Marks
		Subject 1 Mark
		Subject 2 Mark
		•
		•
		Subject N Mark
	Choice List	Total Choice Area Number
		Choice 1
		Choice 2
		•
		Choice N

[SPACE][APPLICATION]". After sending these information as a text message, the server replies with the status of the student. Figure 4 shows the implementation of SMS facility for the interaction with the system. We consider such type of facility due to that fact that most of the students in Bangladesh live in the rural areas where there is not good Internet access facility but there are mobile networks in all most every rural areas of Bangladesh.

Table 14. Administration information

Table Name	Column Family	Column Qualifier
Administration	General Information	Admin Name
		Password

Fig. 4. Interaction with the system via SMS

4   Experiments

We have performed the experiment in a simulation environment of four computers running Ubuntu 64 bit operating system in each of them. The configuration of each computer consists 4 cores, 4 GB RAM and 250 GB SATA disk. At we setup jdk-7 and then setup Hadoop and HBase. We configure Hadoop NameNode, JobTracker, HBase Master, and ZooKeeper on the same node. Here we use Hadoop-1.1.2 and Hbase-0.94.18.

After completing the setup procedure, we create all the necessary tables and define necessary method for accessing the tables. Then, we generate synthetic data and evaluate the performance of our system based on the synthetic data. We compare the performance of our system that has been implemented using HBase with relational database system MySQl.

We first evaluate the insertion performance. Figure 5 shows the insertion performance while using HBase and MySQL. From Figure 5, we can see that when data volume is very small, the performance of HBase and MySQL are almost similar. However, when the data volume increases there is a huge performance degradation in MySQL but there is almost no performance degradation in HBase.

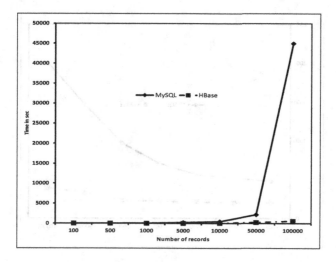

Fig. 5. Comparative insertion performance

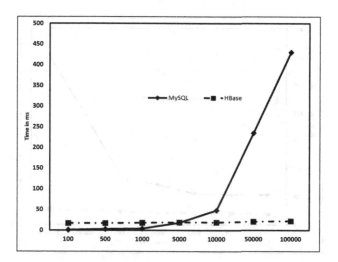

Fig. 6. Comparative query performance

In our next experiment, we measure comparative query performance between HBase and MySQL. The result is shown in Figure 6. From the graph of Figure 6, we can find that for the small data MySQL shows better performance. However, with the increase of data volume, HBase outperforms MySQL by a significant margin.

Figure 7 shows comparative update time. Here, we can find that with the increase of data size, there is an increase in update time in MySQL database. However, there is almost no increase of data update time in HBase.

Our last experiment measures comparative performance of data deletion. Figure 8 shows the results. From the graph of Figure 8, we can see that as data size

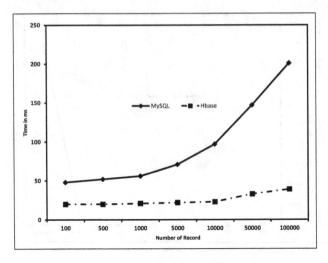

Fig. 7. Comparative update performance

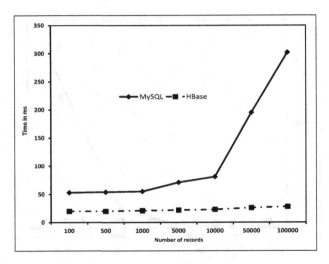

Fig. 8. Comparative deletion performance

increases, the deletion time also increases in MySQL. However, we can find that there is almost no increase of deletion time in HBase.

5   Summary and Conclusions

In this paper, we have focused on the applicability of NoSQL databases for handling moderate volume of data and implemented an application using HBase for a cluster-wise entrance examination system for universities. We performed performance analysis considering an application with synthetic data.

We found that NoSQL databases can be a good choice for applications with moderate load.

References

1. Chang, F., Dean, J., Ghemawat, S., Hsieh, W.C., Wallach, D.A., Burrows, M., Chandra, T., Fikes, A., Gruber, R.E.: Bigtable: A distributed storage system for structured data. In: Proc. of 7th USENIX Symposium on Operating Systems Design and Implementation, pp. 205–208 (2006)
2. DeCandia, G., Hastorun, D., Jampani, M., Kakulapati, G., Lakshman, A., Pilchin, A., Sivasubramanian, S., Vosshall, P., Vogels, W.: Dynamo: Amazon's highly available key-value store. In: Proc. of 21st ACM Symposium on Operating Systems Principles, pp. 205–220 (2007)
3. MongoDB, http://www.mongodb.org
4. Cassandra, http://cassandra.apache.org
5. HBase, http://hbase.apache.org
6. Neo4j, http://neo4j.com/
7. Harter, T., Borthakur, D., Dong, S., Aiyer, A., Tang, L., Arpaci-Dusseau, A.C., Arpaci-Dusseau, R.H.: Analysis of HDFS under HBase: A Facebook messages case study. In: Proc. of 12th USENIX Conference on File and Storage Technologies, pp. 1–14 (2014)
8. Dimiduk, N., Khurana, A.: HBase in action. Manning Publications Co, 3–75 (2013)
9. Kaur, M., Kaur, H.: Concurrency control in distributed database system. International Journal of Advanced Research in Computer Science and Software Engineering 3(7), 1443–1447 (2013)
10. Stevic, M.P., Sad, N.: Serbia: Managing documents with NoSQL in service oriented architecture. Journal of Applied Knowledge Management 1(2), 105–115 (2013)
11. Franke, C., Morin, S., Chebotko, A., Abraham, J., Brazier, P.: Distributed semantic web data management in HBase and MySQL cluster. In: Proc. of the 4th IEEE International Conference on Cloud Computing, pp. 105–112 (2011)
12. Cattell, R.: Scalable SQL and NoSQL data stores. SIGMOD Record 39(4), 12–27 (2010)
13. Pokorny: Databases in the 3rd millennium: trends and research directions. Journal of Systems Integration 1(1), 3–15 (2010)
14. Gajendran, S.K.: A survey on NoSQL databases. Technical report (2012)

A Large Sky Survey Project
and the Related Big Data Analysis

Naoki Yoshida*

Kavli Institute for the Physics and Mathematics of the Universe, WPI,
University of Tokyo
5-1-5 Kashiwanoha, Chiba 277-8583, Japan

Abstract. We explore the frontier of statistical computaional cosmology
using the large imaging data delivered by Subaru Hyper-Suprime-Cam
(HSC) Survey. The large sky usrvey is led by an international group
and will utilizes the 8.3-meter telescope Subaru for 300 nights during
the period of 2014 to 2019. Deep images of a large fraction of sky over
one thousand square degrees will be collected. Our objectives here are
of two folds: we analyse the images of about a half billion galaxies to
reconstruct the distribution of cosmic dark matter, and we detect a few
hundred supernovae that can be used as distance indicators. Combined
together, these two data will enable us to derive fundamental parameters,
the so-called cosmological paramaters, and to predict the evolution of the
universe to the future.

Keywords: astronomy, imaging data.

1 Introduction

Astronomy has been a strong driver of data science. The interaction with data
analysis can be traced back to the 16th century when Tycho Brahe collected an
enourmous amount of data of parallax and brightness measurement (by naked
eye!) of stars and planets. The data were later analysed by Kepler, who then
derived three fundamental laws of the motion of planets. With the advent of
bigger and bigger telescopes, the total size of data collected by the telescopes
has been also increasing dramatically. A good example is the Sloan Digital Sky
Survey, which performed a large systematic survey for over one quarter of the
sky. The total data amounts to 20TB, and of course the scientific outcome and
impact of SDSS is not comparable to any existing survey of similar kind. We are
now entering a new era of Big Data Astronomy. In several to ten years, large
sky-survey projects such as Large Synoptic Survey Telescope (LSST), Square
Kilometre Telescope, and Euclid will commence. For example, LSST is expected
to produce 15 TB per one night(!). Clearly, we will be facing to challenges in
handling the large set of data in a timely manner.

---

* Funded by CREST, Japan Science and Technology Agency.

1.1 Cosmic Dark Matter

An array of recent astronomical observations established the so-called standard cosmological model, in which the energy content of the present-day universe is dominated by mysterious substances called dark matter and dark energy. According to the result, the ordinary matter contribute only 5 percent(!) to the total, and the remaining 95 percent is something yet to be understood. Although we do not know exactly what the two 'dark' substances are, we know the effective roles played by them in terms of the evolution of the universe. Dark energy, giving about 70 percent (hence the largest) contribuion, is thought to fill the space rather uniformly. It causes effectively repulsive force so that the expansion of the universe is accelerated. Dark matter, giving about 25 percent contribution, behaves literally like matter and causes gravitational force mutually and also with the other ordinary matter. Its main role is perhaps to pull together materials to form astronomical objects such as stars and galaxies.

According to Eistein's theory of general relativity predicts, space-time around a massive object is significantly curved and thus light (electromagnetic waves) propagates straight in the curved space time; the resulting spectacular phenomena is called gravitational lensing. The weak version of the phenomena appears as weak lensing, through which images of background galaxies are slightly deformed and aligned to nearby galaxies. There is a rigorous mathematical formulation that provides a method to reconstruct the intervening mass distribution (including dark matter) from collective image distortions. This is exactly what we'd like to do in the next five years using the data from HSC survey.

1.2 Supernovae as Cosmic Standard Candles

Supernovae (SNe) are explosions of stars when they end their lives. There are a variety of types of SNe, depending on the nature of the progenitor star and the explosion mechanism. Here we are interested in a particular type called Type Ia SNe, because they can be used as accurate distance indicators. Distance measurement has always been the central issue in astronomy, because of its intrinsic difficulty to be applied to our vast universe. The notion remains true still in the 21st century, but precision measurement using large telescopes is now possible. It is known since early 1990's that there is a tight correlation between the 'stretch' of the lightcurve of Type Ia SNe, i.e., its typical duration, and the maximum luminosity. The relation perhaps originates from the explosion mechanism, but the details are unknown. However, the important fact is that the intrinsic luminosity of a Type Ia SN can be estimated accurately from observations of its lightcurve. Therefore we can use Type Ia SNe as standard 'candles' to measure the distance to them. Our expectation is that we will detect over ten thousand such supernovae over the next five years using HSC survey.

1.3 HSC Survey and Statistical Computational Cosmology

In 2014, a new wide field camera is installed at the prime focus of Subaru telescope. The Hyper-Suprime-Cam (HSC) has 104 CCDs on the focal plane, and a field of view of about 1.5 square degrees. One snapshot of this camera produces a 1 Giga pixel image. HSC utillizes five broad-band filters in optical to near-infrared wavelength, and also a few narrow-band filters. With the expected 300 nights observation, it will collect literally big data of 25 trillion pixels in total. Our primary goals using the big data are of the following two folds. First, we detect slight distortion of galaxy shapes due to gravitational lensing in order to reconstruct the matter (including dark matter) distribution. Our second purpose is to perform time-differencing of images of the same patch of the sky. We search brightness varitions of some objects and classify them. In fact, many of detected signals are likely false signals, or at least they are not what we look for. There are intrinsic difficulties in this process, as the readers may naively imagine. The process cannot be automated easily, but the amount of data is really demanding us to do automated detection and possibly classification (see, e.g. reference [1,2]). Our team includes researchers in machine learning and we are currently working together to this end.

Suppose we will have successfully reconstructed the large-scale matter distribution and also measured distances to many Type Ia SNe. Then we can use the combined data to accurately measure the growth of large-scale structure and to the cosmic expansion history as a function of time. We will then be able to determine several fundamental parameters in the so-called Friedmann equation and the effective equation-of-state of dark energy. Ultimately, we can *predict* the evolution of cosmic expansion since the Big Bang to the future.

2   Prospects

HSC survey began in 2014, and as of January 2015, it had 10 nights successful observations. By March 2016, data of about 50 nights, which amounts to one sixth of the expected total, will be obtained. It is still too early to perform our statistical study, but the data over 2014-2015 can be used for transient object detections. We are making rapid progress in developping automated transient detection and classification.

Acknowledgments. The author is grateful for financial support by JST CREST.

References

1. Bloom, J.S., et al.: Publications of the Astronomical Society of the Pacific 124, 1175–1196 (2012)
2. Brink, H., et al.: Monthly Notices of the Royal Astronomical Society 435, 1047–1060 (2013)

A Photometric Machine-Learning Method
to Infer Stellar Metallicity

Adam A. Miller[1,2,*]

[1] Jet Propulsion Laboratory, California Institute of Technology,
Pasadena, CA 91109, USA
amiller@astro.caltech.edu
http://astro.caltech.edu/~amiller
[2] California Institute of Technology, Pasadena, CA 91125, USA

Abstract. Following its formation, a star's metal content is one of the few factors that can significantly alter its evolution. Measurements of stellar metallicity ([Fe/H]) typically require a spectrum, but spectroscopic surveys are limited to a few$\times 10^6$ targets; photometric surveys, on the other hand, have detected $> 10^9$ stars. I present a new machine-learning method to predict [Fe/H] from photometric colors measured by the Sloan Digital Sky Survey (SDSS). The training set consists of ~120,000 stars with SDSS photometry and reliable [Fe/H] measurements from the SEGUE Stellar Parameters Pipeline (SSPP). For bright stars ($g' \leq 18$ mag), with $4500\,\mathrm{K} \leq T_\mathrm{eff} \leq 7000\,\mathrm{K}$, corresponding to those with the most reliable SSPP estimates, I find that the model predicts [Fe/H] values with a root-mean-squared-error (RMSE) of ~0.27 dex. The RMSE from this machine-learning method is similar to the scatter in [Fe/H] measurements from low-resolution spectra.

Keywords: photometric surveys, machine learning, random forest, stellar metallicity.

1 Introduction

The Sloan Digital Sky Survey (SDSS, [13]) has cataloged more than one billion photometric sources, while also obtaining nearly 2 million optical spectra [1]. Despite this unprecedented volume of spectra, existing and currently planned instruments have no hope of observing each of the photometrically cataloged stars found by SDSS. Within the next decade, the Large Survey Synoptic Telescope (LSST; [7]) will dwarf SDSS, and other similar surveys, by detecting ~20 billion photometric sources. The data volume from modern photometric surveys is too large to be examined on a source by source basis. Instead, a prudent analysis of the full data set requires advanced algorithms, such that we can identify the most interesting sources for spectroscopic observations, while also inferring the properties of those for which spectra will never be obtained.

---

* NASA Hubble Fellow.

W. Chu et al. (Eds.): DNIS 2015, LNCS 8999, pp. 231–236, 2015.

Machine-learning methods provide a promising solution to this issue: machines can readily identify patterns within the data, enabling a fast classification of the billions of stars detected in modern imaging surveys. One reason machine-learning methods are appealing is that they are data driven: the relationships they derive between observables and the parameters of interest do not rely on parametric physical models. Thus, in scenarios where we are partially ignorant to the relevant stellar physics, the machines may still be able to infer the desired stellar quantities.

Many studies have utilized machine-learning approaches to classify stellar sources of variable brightness (e.g., [4,11,5]), but only recently have efforts been made to infer fundamental physical properties via machine learning [10]. These efforts build on a long history of methods designed to estimate stellar properties, which are typically measured via spectra, from photometric observations. While the effective temperature of a star, T_{eff}, can be photometrically measured with great accuracy [6], estimates of [Fe/H] prove far more challenging [3].

A star with enhanced metal content (i.e. large [Fe/H]) produces less flux in the blue portion of its optical spectrum. Thus, imaging surveys with blue filters, such as SDSS and LSST, can be used to estimate metallicity via the photometric colors of a star. For samples restricted to F and G dwarf stars, broadband colors are capable of producing a scatter $\sim$0.2 dex for [Fe/H] [6]. When no restrictions are applied the best estimates from photometric methods produce a scatter of $\sim$0.3 dex [8].

Here, I present a new machine-learning method, which utilizes the random forest algorithm [2], that is capable of estimating [Fe/H] from the SDSS broadband photometric filters $(u'g'r'i'z')$. I train the model using a sample of $\sim$120,000 stars that have reliable estimates of [Fe/H] from SDSS spectroscopic observations. The final model enables a precise estimate of [Fe/H] with a low catastrophic error rate.

2   Sample

The training set for the machine learning model is constructed from the sample of stars with existing SDSS optical spectra. Every SDSS optical spectrum obtained through the eighth data release was analyzed by the SEGUE Stellar Parameters Pipeline (SSPP), a suite of algorithms optimized to estimate effective temperature (T_{eff}), surface gravity (log g), and metallicity ([Fe/H]) for stellar sources [9]. Briefly, the SSPP provides estimates of these values using multiple methods that are robustly combined to produce final adopted values of T_{eff}, log g, and [Fe/H], as well as their corresponding uncertainties. For high signal-to-noise ratio (SNR) spectra with 4500 K $\leq T_{eff} \leq$ 7500 K and log $g > 2$, the SSPP measures T_{eff}, log g, and [Fe/H] with typical uncertainties of 157 K, 0.29 dex, and 0.24 dex, respectively [9]. The pipeline also flags spectra for which it cannot provide reliable estimates of the stellar parameters.

For the training sample, I include only stars that did not raise any flags during SSPP processing. From this sample of 376,073 stars, I further reject sources with flagged SDSS photometry, a single SSPP measurement of [Fe/H], $T_{eff} < 4500$ K or $T_{eff} > 7000$ K, or $g > 18$ mag. Finally, I remove any duplicate spectroscopic observations of the same star. These cuts are made to ensure that both the photometric and spectroscopic uncertainties are small. I summarize the cuts, as well as the number of stars remaining following each cut below:

(1) SSPP flag = nnnnn   (376,073)
(2) No SDSS photometric flags   (217,274)
(3) 4500 K $\leq T_{eff} \leq$ 7500 K   (188,716)
(4) $\geq$2 SSPP [Fe/H] measurements (182,408)
(5) $g' \leq$ 18 mag   (139,176)
(6) Remove duplicates   (119,596).

In sum, there are ~120,000 stars with reliable photometry and spectroscopic measurements of [Fe/H] that are included in the model training set.

3   Model and Results

There are four features to be utilized by the machine learning model, the SDSS photometric colors ($u' - g'$, $g' - r'$, $r' - i'$, $i' - z'$), which will enable the prediction of [Fe/H], as measured from the SDSS spectra. To perform this supervised machine-learning regression between photometric colors and [Fe/H], I adopt the random forest algorithm [2]. In short, random forest regression aggregates the results of multiple decision trees built from randomized bootstrap samples of the training set. At each node of the individual trees, the splitting parameter is selected from a random subset of the four features in the model to minimize the root-mean-squared-error (RMSE) in the resulting branches from the node. After the forest has been fully constructed, the output from each tree is averaged to provide a robust estimate of [Fe/H].

To optimize the model, the sample of ~120,000 sources is split into a training set containing a random subset of 80,000 stars, while the remaining 39,596 sources provide a test set. Tuning parameters for the random forest are adopted following a grid search and 10-fold cross-validation on the training set. The cross-validated RMSE on the training set is 0.269 dex. The results from applying the optimized model to the test set are shown in Figure 1. The RMSE for the test set is 0.273 dex, and the catastrophic error rate (CER), defined as the percentage of predictions that are incorrect by more than 0.75 dex, is 2.3%. As seen in Figure 1, the model shows a tight scatter around the one-to-one regression line. As noted above, the SSPP produces estimates of [Fe/H] with a typical uncertainty of ~0.24 dex. Thus, this machine learning method produces a scatter similar to that from a low-resolution spectrum. However, with $> 10^9$ SDSS photometrically observed sources and ~3 orders of magnitude fewer spectroscopically observed sources, the machine learning method can be applied to a significantly larger swath of stars.

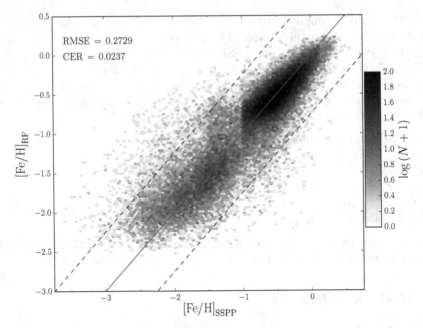

Fig. 1. Final results from the optimized random forest regression model to determine [Fe/H] from SDSS photometric colors. The spectroscopically measured values of [Fe/H] are shown on the abscissa, while the cross-validated random forest predictions are shown on the ordinate. Individual points show the density of sources in a given pixel, as color-coded according to the legend on the right. The overall performance of the model is good, with a cross-validated root-mean-square-error of ∼0.273 dex. The catastrophic error rate is small, with only 2.4% of sources having a predicted metallicity that differs from the spectroscopically measured value by more than 0.75 dex. The solid red line shows the location of a perfect one-to-one regression, while the dashed grey lines show the boundaries for catastrophic prediction errors.

4   Conclusions

Metallicity is a fundamental parameter of all stars. I have demonstrated that for stars with $4500\,\mathrm{K} \leq T_{\mathrm{eff}} \leq 7000\,\mathrm{K}$ and reliable $u'g'r'i'z'$ photometry it is possible to measure [Fe/H] with a typical scatter of ∼0.27 dex. In addition to being reliable, this method is fast and can readily be applied to billions of stars. Thus, it is possible to provide metallicity measurements for a few orders of magnitude more stars than current spectroscopic surveys. The potential applications of the method are numerous, including: the search for stellar structures in the Milky Way (e.g., [6]) or the discovery of the rare class of extremely metal poor stars (e.g., [12]). As additional wide-field photometric surveys come online, machine-learning techniques, such as the one described here, promise to shed light on several mysteries concerning the formation of the Milky Way.

Acknowledgement. I am thankful to Brian Bue and Umaa Rebbapragada for fruitful conversations on model selection. I am grateful for support from a NASA

Hubble Fellowship grant: HST-HF-51325.01, awarded by STScI, operated by AURA, Inc., for NASA, under contract NAS 5-26555. This research was carried out at the Jet Propulsion Laboratory, California Institute of Technology, under a contract with the National Aeronautics and Space Administration.

References

1. Ahn, C.P., Alexandroff, R., Allende Prieto, C., Anders, F., Anderson, S.F., Anderton, T., Andrews, B.H., Aubourg, É., Bailey, S., Bastien, F.A. et al.: The Tenth Data Release of the Sloan Digital Sky Survey: First Spectroscopic Data from the SDSS-III Apache Point Observatory Galactic Evolution Experiment. 211, 17 (April 2014)
2. Breiman, L.: Random forests. Machine Learning 45(1), 5–32 (2001), http://dx.doi.org/10.1023/A
3. Brown, T.M., Latham, D.W., Everett, M.E., Esquerdo, G.A.: Kepler Input Catalog: Photometric Calibration and Stellar Classification. 142, 112 (October 2011)
4. Debosscher, J., Sarro, L.M., Aerts, C., Cuypers, J., Vandenbussche, B., Garrido, R., Solano, E.: Automated supervised classification of variable stars. I. Methodology. 475, 1159–1183 (2007)
5. Dubath, P., Rimoldini, L., Süveges, M., Blomme, J., López, M., Sarro, L.M., De Ridder, J., Cuypers, J., Guy, L., Lecoeur, I., Nienartowicz, K., Jan, A., Beck, M., Mowlavi, N., De Cat, P., Lebzelter, T., Eyer, L.: Random forest automated supervised classification of Hipparcos periodic variable stars 414, 2602–2617 (July 2011)
6. Ivezić, Ž., Sesar, B., Jurić, M., Bond, N., Dalcanton, J., Rockosi, C.M., Yanny, B., Newberg, H.J., Beers, T.C., Allende Prieto, C., Wilhelm, R., Lee, Y.S., Sivarani, T., Norris, J.E., Bailer-Jones, C.A.L., Re Fiorentin, P., Schlegel, D., Uomoto, A., Lupton, R.H., Knapp, G.R., Gunn, J.E., Covey, K.R., Smith, J.A., Miknaitis, G., Doi, M., Tanaka, M., Fukugita, M., Kent, S., Finkbeiner, D., Munn, J.A., Pier, J.R., Quinn, T., Hawley, S., Anderson, S., Kiuchi, F., Chen, A., Bushong, J., Sohi, H., Haggard, D., Kimball, A., Barentine, J., Brewington, H., Harvanek, M., Kleinman, S., Krzesinski, J., Long, D., Nitta, A., Snedden, S., Lee, B., Harris, H., Brinkmann, J., Schneider, D.P., York, D.G.: The Milky Way Tomography with SDSS. II. Stellar Metallicity 684, 287–325 (2008)
7. Ivezić, Ž., Tyson, J.A., Acosta, E., Allsman, R., Anderson, S.F., Andrew, J., Angel, R., Axelrod, T., Barr, J.D., Becker, A.C., Becla, J., Beldica, C., Blandford, R.D., Bloom, J.S., Borne, K., Brandt, W.N., Brown, M.E., Bullock, J.S., Burke, D.L., Chandrasekharan, S., Chesley, S., Claver, C.F., Connolly, A., Cook, K.H., Cooray, A., Covey, K.R., Cribbs, C., Cutri, R., Daues, G., Delgado, F., Ferguson, H., Gawiser, E., Geary, J.C., Gee, P., Geha, M., Gibson, R.R., Gilmore, D.K., Gressler, W.J., Hogan, C., Huffer, M.E., Jacoby, S.H., Jain, B., Jernigan, J.G., Jones, R.L., Juric, M., Kahn, S.M., Kalirai, J.S., Kantor, J.P., Kessler, R., Kirkby, D., Knox, L., Krabbendam, V.L., Krughoff, S., Kulkarni, S., Lambert, R., Levine, D., Liang, M., Lim, K., Lupton, R.H., Marshall, P., Marshall, S., May, M., Miller, M., Mills, D.J., Monet, D.G., Neill, D.R., Nordby, M., O'Connor, P., Oliver, J., Olivier, S.S., Olsen, K., Owen, R.E., Peterson, J.R., Petry, C.E., Pierfederici, F., Pietrowicz, S., Pike, R., Pinto, P.A., Plante, R., Radeka, V., Rasmussen, A., Ridgway, S.T., Rosing, W., Saha, A., Schalk, T.L., Schindler, R.H., Schneider, D.P., Schumacher, G., Sebag, J., Seppala, L.G., Shipsey, I., Silvestri, N., Smith, J.A., Smith, R.C., Strauss, M.A., Stubbs, C.W., Sweeney, D., Szalay, A., Thaler, J.J., Vanden Berk, D., Walkowicz, L., Warner, M., Willman, B., Wittman, D., Wolff, S.C., Wood-Vasey, W.M., Yoachim, P., Zhan, H.: for the LSST Collaboration: LSST: from Science Drivers to Reference Design and Anticipated Data Products. ArXiv e-prints (May 2008)

8. Kerekes, G., Csabai, I., Dobos, L., Trencséni, M.: Photo-Met: A non-parametric method for estimating stellar metallicity from photometric observations. Astronomische Nachrichten 334, 1012 (2013)

9. Lee, Y.S., Beers, T.C., Sivarani, T., Allende Prieto, C., Koesterke, L., Wilhelm, R., Re Fiorentin, P., Bailer-Jones, C.A.L., Norris, J.E., Rockosi, C.M., Yanny, B., Newberg, H.J., Covey, K.R., Zhang, H.T., Luo, A.L.: The SEGUE Stellar Parameter Pipeline. I. Description and Comparison of Individual Methods. 136, 2022–2049 (2008)

10. Miller, A.A., Bloom, J.S., Richards, J.W., Lee, Y.S., Starr, D.L., Butler, N.R., Tokarz, S., Smith, N., Eisner, J.A.: A Machine-learning Method to Infer Fundamental Stellar Parameters from Photometric Light Curves. 798, 122 (2015)

11. Richards, J.W., Starr, D.L., Butler, N.R., Bloom, J.S., Brewer, J.M., Crellin-Quick, A., Higgins, J., Kennedy, R., Rischard, M.: On Machine-learned Classification of Variable Stars with Sparse and Noisy Time-series Data 733, 10 (2011)

12. Schlaufman, K.C., Casey, A.R.: The Best and Brightest Metal-poor Stars. 797, 13 (2014)

13. York, D.G., Adelman, J., Anderson, Jr., J.E., Anderson, S.F., Annis, J., Bahcall, N.A., Bakken, J.A., Barkhouser, R., Bastian, S., Berman, E., Boroski, W.N., Bracker, S., Briegel, C., Briggs, J.W., Brinkmann, J., Brunner, R., Burles, S., Carey, L., Carr, M.A., Castander, F.J., Chen, B., Colestock, P.L., Connolly, A.J., Crocker, J.H., Csabai, I., Czarapata, P.C., Davis, J.E., Doi, M., Dombeck, T., Eisenstein, D., Ellman, N., Elms, B.R., Evans, M.L., Fan, X., Federwitz, G.R., Fiscelli, L., Friedman, S., Frieman, J.A., Fukugita, M., Gillespie, B., Gunn, J.E., Gurbani, V.K., de Haas, E., Haldeman, M., Harris, F.H., Hayes, J., Heckman, T.M., Hennessy, G.S., Hindsley, R.B., Holm, S., Holmgren, D.J., Huang, C.h., Hull, C., Husby, D., Ichikawa, S.I., Ichikawa, T., Ivezić, Ž., Kent, S., Kim, R.S.J., Kinney, E., Klaene, M., Kleinman, A.N., Kleinman, S., Knapp, G.R., Korienek, J., Kron, R.G., Kunszt, P.Z., Lamb, D.Q., Lee, B., Leger, R.F., Limmongkol, S., Lindenmeyer, C., Long, D.C., Loomis, C., Loveday, J., Lucinio, R., Lupton, R.H., MacKinnon, B., Mannery, E.J., Mantsch, P.M., Margon, B., McGehee, P., McKay, T.A., Meiksin, A., Merelli, A., Monet, D.G., Munn, J.A., Narayanan, V.K., Nash, T., Neilsen, E., Neswold, R., Newberg, H.J., Nichol, R.C., Nicinski, T., Nonino, M., Okada, N., Okamura, S., Ostriker, J.P., Owen, R., Pauls, A.G., Peoples, J., Peterson, R.L., Petravick, D., Pier, J.R., Pope, A., Pordes, R., Prosapio, A., Rechenmacher, R., Quinn, T.R., Richards, G.T., Richmond, M.W., Rivetta, C.H., Rockosi, C.M., Ruthmansdorfer, K., Sandford, D., Schlegel, D.J., Schneider, D.P., Sekiguchi, M., Sergey, G., Shimasaku, K., Siegmund, W.A., Smee, S., Smith, J.A., Snedden, S., Stone, R., Stoughton, C., Strauss, M.A., Stubbs, C., SubbaRao, M., Szalay, A.S., Szapudi, I., Szokoly, G.P., Thakar, A.R., Tremonti, C., Tucker, D.L., Uomoto, A., Vanden Berk, D., Vogeley, M.S., Waddell, P., Wang, S.i., Watanabe, M., Weinberg, D.H., Yanny, B., Yasuda, N., SDSS Collaboration: The Sloan Digital Sky Survey: Technical Summary. 120, 1579–1587 (2000)

Query Languages for Domain Specific Information from PTF Astronomical Repository

Yilang Wu[1] and Wanming Chu[2]

[1] Computer Networks Laboratory, Aizu University,
Aizu-Wakamatsu, 965-8580, Japan
d8152103@u-aizu.ac.jp
[2] Database Laboratory, Aizu University,
Aizu-Wakamatsu, 965-8580, Japan
w-chu@u-aizu.ac.jp

Abstract. The increasing availability of vast amount of astronomical repositories on the cloud has enhanced the importance of query language for the domain-specific information. The widely used keyword-based search engines (such as Google or Yahoo), fail to suffice for the needs of skilled/semi-skilled users due to irrelevant returns. The domain specific astronomy query tools (such as Astroquery, CDS Portal, or XML) provide a single entry point to search and access multiple astronomical repositories, however these lack easy query composition tools in unit-step or multi-stages query. Based on the previous research studies on domain-specific query language tools, we aim to implement a query language for obtaining the domain-specific information from the astronomical repositories (such as PTF data).

Keywords: Astronomical Information, Domain-specific Information, Multi-stage Query Language.

1   Introduction

Astronomy is now a data-intensive science. Seeking astronomical information through queries is gaining importance in the astronomical domain. The widely used keyword-based search engines such as Google, Yahoo fail to suffice the needs of the astronomy workers (who are well-versed with the domain knowledge required for querying), who have precise queries and expect complete results within time limits (almost real time). And the existing domain-specific searching tools such as Astroquery[1], CDS Portal[2] or XML[3] have not been fully adopted by the current popular astronomical repositories, and access is still based on keyword based search. It is not easy to use. Thus, in this study, we introduce a multi-stage query language for the domain-specific information from the astronomical repositories, such as Palomar Transient Factory (PTF)[4] data.

The proposed multi-stage query language provides a user-level query calculator to formulate a query using domain concepts. These will simplify the querying tasks for the expert and novice domain users. It will enable them to get the desired results[5].

W. Chu et al. (Eds.): DNIS 2015, LNCS 8999, pp. 237–243, 2015.
© Springer International Publishing Switzerland 2015

2   Background

Astronomy is facing a major data avalanche, from multi-terabyte sky surveys and archives, to billions of detected sources, and hundreds of measured attributes per source[6]. The advent of wide-field synoptic imaging has re-invigorated the venerable field of time domain astronomy[7], which involves the study of "how do the astronomical object change with time". It brings new scientific opportunities and also fresh challenges, including handling a huge amount of data storage and transfer, data mining techniques, classification, and heterogeneous data[8][9].

The data overload breeds query tools. The Astroquery[1] is an Astropy (Python Library for Astronomy) affiliated package that contains a collection of tools to access the big tables of online Astronomical data, being advanced in web service specific interfaces. The Strasbourg astronomical Data Center (CDS) is dedicated to the collection and worldwide distribution of astronomical data and related information through HTML, hosting the SIMBAD astronomical database, VizieR catalogue[10] service, and the Aladin[11] interactive software sky atlas[2]. Meanwhile, to seamlessly utilise the highly specialised astronomical datasets or control systems, the XML, for its rich in semantic definition[12], is adopted for a general and highly extensive framework[3].

Our motivation of introducing easy query for the big scientific data in the time domain astronomy was inspired by the iPTF summer school 2014[13], and reports on the Palomar Transient Factory (PTF)[14]. The PTF is a multi-epochal robotic survey of the northern sky that acquires data[4][15] for the scientific study of transient and variable astrophysical phenomena.

3   Astronomical Data Objects for Query

There are variety of astronomical large data repository publicly availabe for download, searching or query, in various formats of tables, image, HTML, XML, or in the form of Map data.

3.1   PTF Data Archive

The PTF data archive is curated by the NASA/IPAC Infrared Science Archive (IRSA). The PTF Level1 data is the initial release (M81, M44, M42, SDSS Stripe 82, and the Kepler Survey Field) in FITS[16] format, including the Epochal (single exposure) Images, and calibrated Photometric Catalogs. The future releases named Level2 data will expand coverage to the entire northern sky, and will include access to the deep coadds (reference images), their associated catalogs, and will include a searchable photometric catalog database. The majority of the released data is at R-band, with a smaller subset at g-band[4].

There are two methods for retrieving PTF data through the IRSA at IPAC: one is an interactive Graphical User Interface (GUI)[17] which is particularly useful as a data exploration tool; and the other is an API (named as IBE) that uses http syntax, providing low-level, program friendly methods for query, including support for the

IVOA Simple Image Access protocol[18]. The query layer of the IRSA Image Server provides the ability to perform spatial and/or relational queries on image metadata tables, with output in IPAC ASCII table, comma-separated-value (CSV), or tab-separated-value (TSV) format. Queries are performed by appending a query string to a base URL identifying the table to query.

3.2   Astroquery

The Astroquery[1] is a set of tools for querying astronomical web forms and databases. All Astroquery modules are supposed to follow the same API, for instance a simplest form of query can be based on coordinates or object names. Most of the modules have been completed using a common API, such as the SIMBAD[19], VizieR[10], IRSA[15] . query modules. The Astroquery serves data as catalogs, archives, simulation data or some other type data, e.g., line list and atomic/molecular cross section and collision rate service. The Astroquery API is as class method in Python. The methods involve query object by name, query region around a coordinate[20].

3.3   SCP Union

The Supernova Cosmology Project (SCP) "Union2.1"[21][22][23] SN Ia compilation [24] brings together data for 833 SNe, drawn from 19 datasets. Of these, 580 SNe pass usability cuts. All SNe were fit using a single lightcurve fitter (SALT2-1) and uniformly analyzed. The data objects in the SCP Union repository involve figures , cosmology tables, lightcurve data[25]. The figure data illustrates $\Omega_m - \Omega_\wedge$, $\Omega_m - w$, Binned w, Binned ρ obtained from CMB, BAO, and SCP Union2.1 constraints, and also the Binned Hubble Diagram and Residuals. The cosmology tables include the Union Compilation Magnitude vs. Redshift Table, the Covariance Matrix, the Full Table of All SNe, and the CosmoMC Code for Implementing Union Compilation. The Lightcurve Data consists of the SCP HST Cluster Survey Supernova Photometry, the SCP High-z 01 Lightcurve Data and Filters, SNe Summary Table[25].

3.4   XML Based Astronomical Data

The scientific need for a homogenous remote telescope image request system is rapidly escalating as more remote or robotic telescopes are brought to function and scientific programs are created or adapted to use such powerful telescopes. To fit the need, the Remote Telescope Markup Language (RTML)[26] embeds traditional astronomical features such as coordinates and exposure times, and allows for prioritised queue scheduling of telescopes while protecting the telescope operating system. The VOEvent[12] is an international XML standard, defined by the IVOA[27], for transmitting information about a recent astronomical transient, with a view to rapid follow-up, such as Skyalert[28].

3.5   Astronomical Linked Data

Astronomical data artifacts and publications exist in disjointed repositories. The conceptual relationship that links data and publications is rarely made explicit.

Seamless Astronomy Group[29][30][31] at the Harvard-Smithsonian Center for Astrophysics tries to let the connections between literature and data grow more seamless and invisible, so that a researcher can spend more time thinking about science, and less about finding information.

4   Query Language Interfaces

Scientific data within the web data resources is often represented in XML or a related form. XML serves an important role. It has been adopted as the standard language for representing structured data for the traditional Web resources. Thus, many Web-based knowledge management repositories store data and documents in XML. Further, the semantics about the data can be represented by modeling these, with an ontology. Then, it is possible to extract knowledge[32]. Ontologies play an important role in realizing the semantic Web, wherein data will be more sharable because their semantics will be represented in Web-accessible ontologies. Recent reports implement an architecture for this ontology using de facto languages of the semantic Web including OWL and RuleML, thus preparing the ontology for use in data sharing[32]. The users of data resouces are often not skilled in the use of programming languages. These users differ from the Web users and database users. Most of the existing document repositories on the Web have alphabetical and keyword based searches. These are not sufficient for the expert users with precise and complex queries, who require in-depth results within time constraints. Their information needs can be supported by providing user-level schema. Such a schema can support database-style high-level query languages over these repositories. Seeking specialized domain-specific information through queries is gaining importance[5].

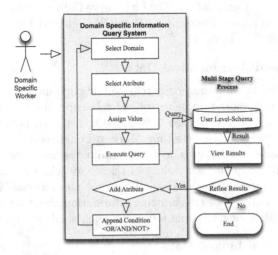

Fig. 1. Query Formation via the Proposed Muti-stage Query Language on User Level

The aim of the is to develop a domain-specific multi-stage query language described in [33] for the archival portals in the scientific domain. Figure 1, represents the query formulation process for astronomical queries, over astronomical objects, through the proposed multi-stage query language. At each stage of query formulation the user can dynamically select an object or a concept to query. Assign a value for it and then either execute the query or further refine the query by adding another attribute(s) and view results. The query is executed on the user-level schema. It provides the users with the segment-level results. Hence, the proposed query language can allow the user to formulate complex DB-style queries using a simple interface and understandable attributes. Figure 1, shows the steps in a typical query formulation process.

5 Summary and Conclusions

A proposed query language for PTF is capable of performing multi-stage visual query (simple, middle, complex, and recursive). We aim to demonstrate a procedure about — how to overcome the existing shortcomings by: (1) implementing the query language to seek diagnostic or hypothesis-directed information (followed by a astronomical domain-expert) and (2) presenting the relevant areas (granular results) of catalog tables or web documents that match the user's query criteria.

In order to achieve the desired query language interfaces for in depth query by skilled and semi-skilled users, it is necessary to organise data as objects in a new schema. Similarly, a set of user level operations, need to be supported. These operations work in a single step at a time. These can be applied in a sequence (by the users) to achieve programmable query language results.

References

1. Ginsburg, A., Robitaille, T., Parikh, M., Deil, C., Mirocha, J., Woillez, J., Svoboda, B., Willett, K., Allen, J.T., Grollier, F., Persson, M.V., Shiga, D.: Astroquery v0.1. 09 (2013)
2. CDS Portal, http://cdsportal.u-strasbg.fr (accessed January 2015)
3. Ames, T., Koans, L., Sall, K., Warsaw, C.: Using XML and Java for telescope and instrumentation control. Advanced Telescope and Instrumentation Control Software 4009, 2–12 (2000)
4. Laher, R.R., Surace, J., Grillmair, C.J., Ofek, E.O., Levitan, D., Sesar, B., van Eyken, J.C., Law, N.M., Helou, G., Hamam, N., Masci, F.J., Mattingly, S., Jackson, E., Hacopeans, E., Mi, W., Groom, S., Teplitz, H., Desai, V., Hale, D., Smith, R., Walters, R., Quimby, R., Kasliwal, M., Horesh, A., Bellm, E., Barlow, T., Waszczak, A., Prince, T., Kulkarni, S.R.: IPAC Image Processing and Data Archiving for the Palomar Transient Factory (April 2014)
5. Madaan, A., Chu, W.: Handling domain specific document repositories for application of query languages. In: Madaan, A., Kikuchi, S., Bhalla, S. (eds.) DNIS 2014. LNCS, vol. 8381, pp. 152–167. Springer, Heidelberg (2014), http://dx.doi.org/10.1007/978-3-319-05693-7_10
6. Djorgovski, S.G.: New Astronomy With a Virtual Observatory Astronomy is Facing a Major

7. Kasliwal, M.: Transients in the Local Universe: Today and Tomorrow. In: International Conference on Computational Physics, Singapore (2015)
8. Graham, M.J., Djorgovski, S.G., Mahabal, A., Donalek, C., Drake, A., Longo, G.: Data challenges of time domain astronomy. Distributed and Parallel Databases 30, 371–384 (2012)
9. Madaan, A., Chu, W., Bhalla, S.: VisHue: Web page segmentation for an improved query interface for medlinePlus medical encyclopedia. In: Kikuchi, S., Madaan, A., Sachdeva, S., Bhalla, S. (eds.) DNIS 2011. LNCS, vol. 7108, pp. 89–108. Springer, Heidelberg (2011), http://dx.doi.org/10.1007/978-3-642-25731-5_9
10. Ochsenbein, F., Bauer, P., Marcout, J.: The VizieR database of Astronomical Catalogues. 10, 10 (2000)
11. Franke, B.: An Introduction To The Aladin Sky Atlas. Tech. rep.
12. Seaman, R., Williams, R., Optical, N., Observatory, A., Allan, A., Barthelmy, S., Bloom, J.S., Brewer, J.M., Denny, R.B., Fitzpatrick, M., Graham, M., Gray, N., Hessman, F., Marka, S., Rots, A., Vestrand, T., Wozniak, P.: VOEvent reporting metadata. pp. 1–27 (2011)
13. iptf summer school, http://phares.caltech.edu/iptf/iptf_SummerSchool_2014/ (accessed august 2014)
14. Law, N.M., Kulkarni, S.R., Dekany, R.G., Ofek, E.O., Quimby, R.M., Nugent, P.E., Surace, J., Grillmair, C.C., Bloom, J.S., Kasliwal, M.M., Bildsten, L., Brown, T., Cenko, S.B., Ciardi, D., Croner, E., Djorgovski, S.G., van Eyken, J.C.: a. V. Filippenko, D. B. Fox, a. Gal-Yam, D. Hale, N. Hamam, G. Helou, J. R. Henning, D. a. Howell, J. Jacobsen, R. Laher, S. Mattingly, D. McKenna, a. Pickles, D. Poznanski, G. Rahmer, a. Rau, W. Rosing, M. Shara, R. Smith, D. Starr, M. Sullivan, V. Velur, R. S. Walters, and J. Zolkower, "The Palomar Transient Factory: System Overview, Performance and First Results," p. 12 (2009)
15. "IRSA Image Server", http://irsa.ipac.caltech.edu/ibe/index.html (accessed: January 2015)
16. Pence, W.D., Chiapetti, L., Page, C.G., Shaw, R.A., Stobie, E.: Definition of the Flexible Image Transport System (FITS), Version 3.0. vol. 42 (2010)
17. "PTF GUI", http://irsa.ipac.caltech.edu/applications/ptf/ (accessed: January 2015)
18. Access, S.I., Tody, D., Dowler, P., Dowler, P., Tody, D., Plante, R.: IVOA Simple Image Access IVOA Proposed Recommendation 2014-07-07, 1–25 (2014)
19. SIMBAD: Set of Identifications, Measurements and Bibliography for Astronomical, http://simbad.u-strasbg.fr/Simbad (accessed: January 2015)
20. Astroquery, https://astroquery.readthedocs.org (accessed: January 2015)
21. Suzuki, N., Rubin, D., Lidman, C., Aldering, G., Amanullah, R., Barbary, K., Barrientos, L.F., Botyanszki, J., Brodwin, M., Connolly, N., Dawson, K.S., Dey, A., Doi, M., Donahue, M., Deustua, S., Eisenhardt, P., Ellingson, E., Faccioli, L., Fadeyev, V., Fakhouri, H.K., Fruchter, A.S., Gilbank, D.G., Gladders, M.D., Goldhaber, G., Gonzalez, A.H., Goobar, A., Gude, A., Hattori, T., Hoekstra, H., Hsiao, E., Huang, X., Ihara, Y., Jee, M.J., Johnston, D., Kashikawa, N., Koester, B., Konishi, K., Kowalski, M., Linder, E.V., Lubin, L., Melbourne, J., Meyers, J., Morokuma, T., Munshi, F., Mullis, C., Oda, T., Panagia, N., Perlmutter, S., Postman, M., Pritchard, T., Rhodes, J., Ripoche, P., Rosati, P., Schlegel, D.J., Spadafora, A., Stanford, S.A., Stanishev, V., Stern, D., Strovink, M., Takanashi, N., Tokita, K., Wagner, M., Wang, L., Yasuda, N., Yee, H.K.C.: The Hubble Space Telescope Cluster Supernova Survey: V. Improving the Dark Energy Constraints Above $z¿1$ and Building an Early-Type-Hosted Supernova Sample 27 (2011)

22. Amanullah, R., Lidman, C., Rubin, D., Aldering, G., Astier, P., Barbary, K., Burns, M.S., Conley, A., Dawson, K.S., Deustua, S.E., Doi, M., Fabbro, S., Faccioli, L., Fakhouri, H.K., Folatelli, G., Fruchter, A.S., Furusawa, H., Garavini, G., Goldhaber, G., Goobar, A., Groom, D.E., Hook, I., Howell, D.A., Kashikawa, N., Kim, A.G., Knop, R.A., Kowalski, M., Linder, E., Meyers, J., Morokuma, T., Nobili, S., Nordin, J., Nugent, P.E., Ostman, L., Pain, R., Panagia, N., Perlmutter, S., Raux, J., Ruiz-Lapuente, P., Spadafora, A.L., Strovink, M., Suzuki, N., Wang, L., Wood-Vasey, W.M., Yasuda, N.: Spectra and Light Curves of Six Type Ia Supernovae at 0.511 ¡ z ¡ 1.12 and the Union2 Compilation, 33 (2010)

23. Kowalski, M., Rubin, D., Aldering, G., Agostinho, R.J., Amadon, A., Amanullah, R., Balland, C., Barbary, K., Blanc, G., Challis, P.J., Conley, A., Connolly, N.V., Covarrubias, R., Dawson, K.S., Deustua, S.E., Ellis, R., Fabbro, S., Fadeyev, V., Fan, X., Farris, B., Folatelli, G., Frye, B.L., Garavini, G., Gates, E.L., Germany, L., Goldhaber, G., Goldman, B., Goobar, A., Groom, D.E., Haissinski, J., Hardin, D., Hook, I., Kent, S., Kim, A.G., Knop, R.A., Lidman, C., Linder, E.V., Mendez, J., Meyers, J., Miller, G.J., Moniez, M., Mourao, A.M., Newberg, H., Nobili, S., Nugent, P.E., Pain, R., Perdereau, O., Perlmutter, S., Phillips, M.M., Prasad, V., Quimby, R., Regnault, N., Rich, J., Rubenstein, E.P., Ruiz-Lapuente, P., Santos, F.D., Schaefer, B.E., Schommer, R.A., Smith, R.C., Soderberg, A.M., Spadafora, A.L., Strolger, L.G., Strovink, M., Suntzeff, N.B., Suzuki, N., Thomas, R.C., Walton, N.A., Wang, L., Wood-Vasey, W.M., Yun, J.L.: Improved Cosmological Constraints from New, Old and Combined Supernova Datasets 49 (2008)

24. Supernova Cosmology Project Union 2.1 Compilation, http://supernova.lbl.gov/union/ (accessed: January 2015)

25. Data Description for Supernova Cosmology Project, http://supernova.lbl.gov/union/descriptions.html (accessed: January 2015)

26. Pennypacker, C., Boer, M., Denny, R., Hessman, F.V., Aymon, J., Duric, N., Gordon, S., Barnaby, D., Spear, G., Hoette, V.: RTML - a standard for use of remote telescopes. Astronomy and Astrophysics 395(2), 727–731 (2002)

27. Documents and Standards about Virtual Observatory, http://www.ivoa.net/documents/index.html (accessed: January 2015)

28. Williams, R.D., Djorgovski, S.G., Drake, A.J., Graham, M.J., Mahabal, A.: Skyalert: Real-time Astronomy for You and Your Robots XXX, 4 (2009)

29. Alyssa Goodmans, Seamless Astronomy, tech. rep. (2013)

30. Seamless Astronomy, http://projects.iq.harvard.edu/seamlessastronomy (accessed: January 2015)

31. Goodman, A., Fay, J., Muench, A., Pepe, A., Udomprasert, P., Wong, C.: World-Wide Telescope in Research and Education. eprint arXiv:1201.1285, 4 (2012)

32. Kim, H., Sengupta, A.: Extracting knowledge from xml document repository: a semantic web-based approach. Information Technology and Management 8(3), 205–221 (2007)

33. Madaan, A., Bhalla, S.: Domain specific multistage query language for medical document repositories. Proc. VLDB Endow. 6(12), 1410–1415 (2013)

Pariket: Mining Business Process Logs for Root Cause Analysis of Anomalous Incidents

Nisha Gupta[1], Kritika Anand[1], and Ashish Sureka[2]

[1] Indraprastha Institute of Information Technology, Delhi (IIITD), India
{nisha1345,kritika1339}@iiitd.ac.in
http://www.iiitd.ac.in/
[2] Software Analytics Research Lab (SARL), India
ashish@iiitd.ac.in
http://www.software-analytics.in/

Abstract. Process mining consists of extracting knowledge and action-able information from event-logs recorded by Process Aware Information Systems (PAIS). PAIS are vulnerable to system failures, malfunctions, fraudulent and undesirable executions resulting in anomalous trails and traces. The flexibility in PAIS resulting in large number of trace variants and the large volume of event-logs makes it challenging to identify anomalous executions and determining their root causes. We propose a framework and a multi-step process to identify root causes of anomalous traces in business process logs. We first transform the event-log into a sequential dataset and apply Window-based and Markovian techniques to identify anomalies. We then integrate the basic eventlog data consisting of the Case ID, time-stamp and activity with the contextual data and prepare a dataset consisting of two classes (anomalous and normal). We apply Machine Learning techniques such as decision tree classifiers to extract rules (explaining the root causes) describing anomalous trans-actions. We use advanced visualization techniques such as parallel plots to present the data in a format making it easy for a process analyst to identify the characteristics of anomalous executions. We conduct a trian-gulation study to gather multiple evidences to validate the effectiveness and accuracy of our approach.

Keywords: Anomalous Incidents, Business Process Mining, Decision Tree Classifier, Event Log, Markovian Based Technique, Root Cause Analysis.

1 Research Motivation and Aim

Business Process Management Systems (BPMS), Workflow Management Systems (WMS) and Process Aware Information Systems (PAIS) log events and activities during the execution of a process. Process Mining is a relatively young and emerging discipline consisting of analyzing the event logs from such systems for extracting knowledge such as the discovery of runtime process model (discovery), checking and verification of the design time process model with the

W. Chu et al. (Eds.): DNIS 2015, LNCS 8999, pp. 244–263, 2015.
© Springer International Publishing Switzerland 2015

runtime process model (conformance analysis) and improving the business process (recommendation and extension) [1]. A process consists of cases or incidents. A case consists of events. Each event in the event log relates to precisely one case. Events within a case are ordered and have attributes such as activity, timestamp, actor and several additional information such as the cost. The incidents and activities in event logs can be modeled as sequential and time-series data. Anomaly detection in business process logs is an area that has attracted several researcher's attention [2] [8]. Anomalies are patterns in data that do not conform to a well defined notion of normal behavior. Anomaly detection in business process logs has several applications such as fraud detection, identification of malicious activity and breakdown of the system and understanding the causes of process errors. Due to complex and numerous business processes in a large organizations, it is difficult for any employee to monitor the whole system. As a consequence of this anomalies occurring in a system remains undetected until serious losses are caused by it. Therefore, Root Cause Analysis (RCA) is done to identify root causes and sources of problems and improve or correct the given process so that major problems can be avoided in future.

The focus of the study presented in this paper is on anomaly detection in business process logs and identification of their root causes. We present a different and fresh perspective to the stated problem and our work is motivated by the need to extend the state-of-the-art in the field of techniques for anomaly detection and RCA in business process event logs. While there has been work done in the area of anomaly detection and RCA in business process logs, to the best of our knowledge, the work presented in this paper is the first focused study on such a dataset for the application of anomaly detection and RCA. The research aim of the work presented in this paper is the following:

1. To investigate Window based and Markovian based techniques for detecting anomalies in business process event logs.
2. To apply machine learning techniques such as decision tree classifier to extract rules describing cause of anomalous behavior.
3. To interactively explore different patterns of data using advanced visualization techniques such as parallel plot.
4. To investigate solutions assisting a process analyst to analyze decision tree and parallel plot results, thus identifying root cause of anomalous incidents.
5. To demonstrate the effectiveness of our proposed approach using triangulation study[1]. We conduct experiments on a recent, large and real-world incident management data of an enterprise.

2   Related Work and Research Contributions

We conduct a literature review of papers closely related to the work presented in this paper. Calderón-Ruiz et al. propose a novel technique to identify potential

---

[1] http://en.wikipedia.org/wiki/Triangulation_(social_science)

causes of failures in business process by extending available Process Mining techniques [7]. They test their technique using several synthetic event logs and are able to successfully find missing or unnecessary activities, and failed behavioural patterns that differ from successful patterns either in the control flow or in the time perspective [7]. Heravizadeh et al. propose a conceptual methodology of root-cause analysis in business processes, based on the definition of softgoals (nonfunctional requirements) for all process activities, as well as correlations between these softgoals and related quality metrics [10]. Suriadi et al. propose an approach to enrich and transform process-based logs for Root Cause Analysis based on classification algorithms [13]. They use decision trees to identify the causes of overtime faults [13]. Vasilyev et al. develop an approach to find the cause of delays based on the information recorded in an event log [14]. The approach is based on a logic representation of the event log and on the application of decision tree induction to separate process instances according to their duration [14].

Bezerra et al. present some approaches based on incremental mining [15] for anomaly detection, but these algorithms cannot deal with longer traces and/or logs with various classes of traces [2]. Then, in order to deal with such constraints, they begin to develop other solutions based on process mining algorithms available in ProM[2] framework [3] [6]. Bezerra et al. propose an anomaly detection model based on the discovery of an "appropriate process model" [6]. Bezerra et al. apply the process discovery and conformance algorithms from ProM framework for implementing the anomaly detection algorithms [3]. Bezerra et al. present three new algorithms (threshold, iterative, and sampling) to detect "hard to find" anomalies in a process log based only on the control-flow perspective of the traces [5]. This work does not deal with anomalous executions of processes that follow a correct execution path but deal with unusual data, or are executed by unusual roles or users, or have unusual timings [5]. Bezerra et al. develop an algorithm more efficient than the Sampling Algorithm [4]. They propose an approach for anomaly detection which is an extension of the Threshold Algorithm also reported in [3] [5], which uses process mining tools for process discovery and process analysis for supporting the detection [4].

In context to existing work, the study presented in this paper makes the following novel contributions:

1. Detection of anomalous traces in business process event-logs using Window-based and Markovian-based techniques (after transforming the event-log into a sequential) dataset.
2. Root Cause Analysis (RCA) of anomalous traces using parallel coordinate plots. Application of parallel coordinate plots for visualizing the characteristics of anomalous and normal traces (representing the traces and their attribute values as a polyline with vertices on the parallel axes).
3. Application of tree diagrams as a visual and analytical decision support tool for identifying the features of anomalous traces, thereby assisting a process analyst in problem solving and Root Cause Analysis (RCA).

---

[2] ProM is a pluggable and open-source framework for Process Mining.

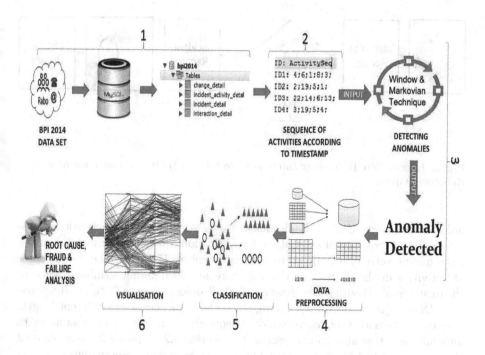

Fig. 1. Architecture diagram and data processing pipeline for Pariket (Mining Business Process Logs for Root Cause Analysis of Anomalous Incidents)

4. An in-depth and focused empirical analysis on a real-world dataset (Rabobank Group[3]: Activity log for incidents) demonstrating the effectiveness of the proposed approach. Application of triangulation technique to validate the outcome of RCA through cross-verification.

3   Research Framework and Solution Approach

Figure 1 shows the high-level architecture diagram of the proposed solution approach (called as *Pariket*). The proposed approach is a multi-step process primarily consists of 6 phases: experimental dataset collection, sequential dataset conversion, anomaly detection, data pre-processing, classification and visualization. The six phases are labeled in the architecture diagram in Figure 1. In phase 1, we download large real world data from Rabobank Group (refer to Section 4 on experimental dataset). The dataset consists of event logs from interactions records, incidents records, incident activities and change records. We choose incidents from incident activities to find out anomalous incident patterns. In phase 2, for a particular incident we order the type of activities according to increasing

---

[3] http://data.3tu.nl/repository/uuid:c3e5d162-0cfd-4bb0-bd82-af5268819c35

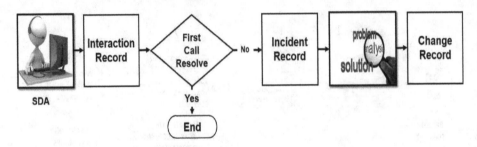

Fig. 2. Information Technology Infrastructure Library (ITIL) process implemented in Rabobank Group

order of DateTime Stamp. Each incident consisting of several activities is represented as a sequence of symbols (refer to Section 5.1 on experimental results). Each unique activity is mapped to a integer symbol. There are 39 different kinds of activities in the dataset and hence there are 39 different symbols. Some of the example of activities are: Referred (28), Problem Closure (22), OOResponse (18), Dial-In (10) and Contact Change (8). The sequences are of different length. The incidents with their corresponding sequence of activities serve as input to anomaly detection algorithms described in Section 5.2. In phase 3, we implement Window based and Markovian based technique [8] to detect anomalous incidents (refer to Section 5.2 on experimental results). We receive top N anomalous incidents as output from anomaly detection algorithms. We apply decision tree classifier and visualization techniques to identify root causes of anomalous incidents. Input to these techniques requires data to be in a particular format and of high quality. Hence, in phase 4 we perform data pre-processing to bring the data in the required format (refer to Section 5.3 of experimental results). In phase 5, we create decision tree using $J48$ algorithm in Waikato Environment for Knowledge Analysis (Weka)[4] (refer to Section 5.4). The $J48$ tree classifier is the $C4.5$ implementation available in Weka. $J48$ handles both numeric and nominal attribute values. In phase 6, we apply advanced visualization technique such as parallel plot in Tibco Spotfire[5] to interactively explore different regions of data (refer to Section 5.5). Business process analyst then analyze decision tree and parallel plot results to identify the root cause of anomalous incidents.

4   Experimental Dataset

We conduct our study on a large real world data from Rabobank Group Information and Communication Technology (ICT). The data is related to the Information Technology Infrastructure Library (ITIL) process implemented in the Bank. The ITIL process depicted in Figure 2 starts when an internal client

---

[4] http://www.cs.waikato.ac.nz/ml/weka/

[5] http://spotfire.tibco.com/

reports an issue regarding disruption of ICT service to Service Desk Agent (SDA). SDA records the complete information about the problem in Interaction Record. If the issue does not get resolved on first contact then a Incident record is created for the corresponding Interaction else the issue is closed. There can be many to one mapping between Interaction Record and Incident Record. If a issue appears frequently then a request for change is initiated.

The dataset is provided in the CSV format. It contains the event logs from interactions records, incidents records, incident activities and change records. The provided dataset is of six month duration from October 2013 - March 2014. Interactions that were not resolved before 31 March, were removed from the dataset. The attributes of original .CSV files are converted to appropriate data types, such as standardized timestamp formats, for analysis. After loading the data on to MySQL database, we build four tables: Interaction_detail, Incident_detail, Incident_activity_detail and Change_detail.

1. Interaction_detail - It has 147,004 records, each one corresponding to an interaction. Every record contains information like InteractionID, Priority, Category, Open Time, Close Time, Handle Time, and First Call Resolution (whether SDA was able to resolve the issue on first contact or not).
2. Incident_detail - It has 46,606 records, each one corresponding to an incident case. Every record has attributes like IncidentID, Related Interaction, Priority, Open Time, Handle Time, Configuration Item Affected etc.
3. Incident_activity_detail - It has 466,737 records. Each record contains an IncidentID with the activities performed on it. It also contains information about the Assignment Group that is responsible for a particular activity.
4. Change_detail - It contains records of the activities performed on each change case. It has information about Configuration Item Affected, Service Component Affected, Change Type and Risk Assessment etc.

As an academic, we believe and encourage academic code or software sharing in the interest of improving *openness and research reproducibility*. We release our code and dataset in public domain so that other researchers can validate our scientific claims and use our tool for comparison or benchmarking purposes (and also reusability and extension). Our code and dataset is hosted on GitHub[6] which is a popular web-based hosting service for software development projects. We select GPL license (restrictive license) so that our code can never be closed-sourced.

5 Experimental Results

We perform a series of steps to identify the root cause of anomalous incidents. Each of the following 6 sub-sections describes the steps consisting of procedure or approach and findings.

---

[6] https://github.com/ashishsureka/pariket

5.1   Sequential Dataset Conversion

We analyze all the tables and amongst them we choose Incident_activity_detail table to find out the anomalous incident patterns. This table contains a log of activities performed by the service team(s) to resolve incidents which are not resolved by first contact. The main reason for choosing this table is because it has information regarding the type of activities performed on a particular incident id and also the timestamp when this incident activity type started to resolve the issue.

The attribute IncidentActivity_Type represents the type of activity performed on the incident. There are 39 unique activities. Some of the examples are: Assignment (ASG), Status Change (STC), Update (UPD), Referred (REF), Problem Closure (PC), OOResponse (OOR), Dial-In (DI) and Contact Change (CC). Figure 3 represents the pareto chart showing the distribution of activities and their cumulative count. The Y-axis is in logarithmic scale. We assign integer number starting with 0 to 38 to these activities, and then we add an extra column IncidentActivity_Type_Number into the table Incident_activity_detail denoting this activity number. For a particular incident we order the activities according to increasing order of DateTime Stamp. This is done for all the unique incidents in the Incident_activity_detail table.

We apply Window Based and Markovian Based Techniques for detecting anomalous incidents. The input dataset to these algorithms has to be in sequential format. Therefore, to accomplish this we create a new table 'IncidentActivitySequence' containing two attributes 'IncidentID' and 'IncidentAcitvity_Type_Number'. Each record in this table contains all unique IncidentID's and sequences of IncidentActivity_Type_Number separated by semicolon according to timestamp from Incident_activity_detail. For example, corresponding to IncidentID 'IM0000012', the sequence of activities are '34;27;2;34;4;5;'.

5.2   Anomaly Detection

The outcome of the Section 5.1 is a list of all incident id's with their corresponding sequence of activities ordered according to timestamp. The aim of algorithms described in this Subsection is to identify anomalous incidents based on the obtained discrete sequences. There is no reference or training database available containing only normal sequences. Hence, our task is to detect anomalous sequences from an unlabeled database of sequences. The problem is of unsupervised anomaly detection. A formal representation of the problem is [8]: Given a set of n sequences, $S = \{S_1, S_2, ...,S_n\}$, find all sequences in S that are anomalous with respect to rest of S. This unsupervised problem can be solved by using a semi-supervised approach where we treat the entire dataset as training set and then score each sequence with respect to this training set. We assume that majority of sequences in the unlabeled database are normal as anomalies are generally infrequent in nature [8]. We use two algorithms, Window Based and Markovian Based described in the following Subsections for anomaly detection.

Algorithm 1: Window Based Algorithm (ID, S, k, λ, N)

Data: IncidentID (ID = $ID_1....ID_n$) and Sequence of activities (S = $S_1...S_n$) from table.

Result: Top N Anomalous IncidentID.

1 set windowSize = k, threshold = λ;
2 create a empty dictionary D, D'
3 create an arrayList anomalousIncidents;
4 **foreach** IncidentID ID_i in ID **do**
5   S_i = get the sequence corresponding to ID_i
6   set windowCount = S_i.length - windowSize + 1
7   **foreach** j = 1 to windowCount **do**
8   read the subsequence (W_j) of length = windowSize starting from j^{th} position in S_i
9   **if** W_j is not present in D **then**
10   add (W_j, 1) as (key, value) pair in D
11   **else**
12   add (W_j , value + 1) in D

13 **foreach** IncidentID ID_i in ID **do**
14   S_i = get the sequence corresponding to ID_i
15   set windowCount = S_i.length - windowSize + 1
16   set anomalyScore = 0.0
17   **foreach** j = 1 to windowCount **do**
18   read the subsequence (W_j) of length = windowSize starting from j^{th} position in S_i
19   get the (key, value) pair from D corresponding to key = W_j
20   **if** value is less than threshold **then**
21   anomalyScore = anomalyScore + 1

22   anomalyScore = anomalyScore / windowCount
23   add ID_i, anomalyScore into D'

24 sort D' according to decreasing anomalyScore
25 add top N IncidentId into anomalousIncidents
26 return anomalousIncidents

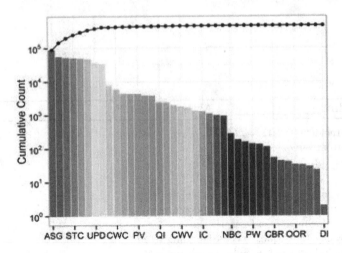

Fig. 3. Pareto chart showing the distribution of activities and their cumulative count. Y-axis is in logarithmic scale.

Window Based Technique. The motivation behind using window based technique is to determine anomalous sequences even if the cause of anomaly is localized to one or more shorter subsequences within the actual sequence [9]. Window based technique in general operates as, first we extract overlapping windows of fixed length (k) from a given test sequence. Then, we assign some anomaly score to each extracted window based on a threshold value (λ). Finally, the anomaly score of all the windows are combined to obtain an anomaly score for the test sequence [8].

The pseudocode for Window Based anomaly detection algorithm is shown in Algorithm 1. The input to the algorithm is data comprising of IncidentID, sequence of activities from table IncidentActivitySequence, window size (k), threshold (λ) and number of anomalous incidents (N). The algorithm returns top N anomalous IncidentID as output. The main challenge was to find out the size of window (k) and the value of threshold (λ). We analyze all the subsequences of window length less than 3. Our analysis reveals that they occur very frequently. Therefore, we cannot take them as anomalous subsequences because according to our previous assumption in Section 5.2 anomalies in our dataset are in minority. Therefore, k has to be equal to or greater than 3.

Algorithm 1 consists of two phases: training and testing. We choose 3 experimental parameters: $k = 3$, $\lambda = 4$ and $N = 1000$. The training phase is represented by Steps 4-12. During this phase, we obtain the sequence of activities for each IncidentID. From the sequence we extract k length overlapping (sliding) windows. We maintain each unique window with its frequency in normal dictionary D. The testing phase is represented by Steps 13-25. Every sequence of the training dataset is considered as the test sequence. During this phase, we extract sliding windows of length k from the test sequence S_i. A window W_j is assigned an anomalyScore of 1 if the frequency associated with the window W_j in dictionary D is less than the threshold value (λ) else anomalyScore is 0. To calculate the anomalyScore of a complete test sequence S_i, we take summation of anomalyScore of all the subsequence windows contained in it. The anomaly score of the test sequence is proportional to the number of anomalous windows in the test sequence [11]. The result obtained after executing Steps 14-21 is then divided by the number of windows contained in the test sequence. This normalization is done to take into account the varying lengths of sequences. The anomalyScore of 1 for a test sequence denotes most anomalous and 0 as least anomalous. We store the IncidentID and its corresponding anomalyScore in the dictionary D'. Then, we sort the IncidentID's in decreasing order of anomalyScore and return top N anomalous IncidentID's.

Markovian Based Technique. We apply fixed Markovian technique [8] which is based on the property of short memory of sequences. This property states that the conditional probability of occurrence of a symbol s_i is dependent on the occurrence of previous k symbols with in a sequence S_i [12]. The conditional probability of occurrence of a symbol s_i in a sequence S_i is given by Equation 1:

$$P(s_i|s_{(i-k)}...s_{(i-1)}) = \frac{freq(s_{(i-k)}...s_i)}{freq(s_{(i-k)}...s_{(i-1)})} \tag{1}$$

where $freq(s_{(i-k)}...s_i)$ is the frequency of occurrence of the subsequence $s_{(i-k)}$ ...s_i in the sequences in S and $freq(s_{(i-k)}...s_{(i-1)})$ is the frequency of occurrence of the subsequence $s_{(i-k)}...s_{(i-1)}$ in the sequences in S.

The pseudocode for Markovian based anomaly detection algorithm is shown in Algorithm 2. The input to the algorithm is data comprising of IncidentID, sequence of activities from table IncidentActivitySequence, window size (k) and number of anomalous incidents (N). The algorithm returns top N anomalous IncidentID as output. Algorithm 2 consists of two phases: training and testing. Steps 4-17 represents the training phase. During this phase, we create two dictionaries D_k and D_{k+1} of length k and $k+1$ respectively. The process for creation of dictionary is similar to that described in Section 5.2. We choose $k = 3$ for our experiment. It takes into account the subsequences of length 4 which are dependent on previous 3 symbols. Steps 18-32 represents the testing phase. Steps 18-32 are repeated for each IncidentID in table IncidentActivitySequence. We extract the test sequence S_i corresponding to a IncidentID ID_i in Step 19 and calculate the number of subsequences of length k in Step 20. Steps 23-28 are repeated for each each subsequence within the test sequence S_i. Step 23 and 24 reads the subsequence W_j and W_{j+1} of length k and $K+1$ respectively starting from position j. Step 25 and 26 calculates the frequency $value_j$ and $value_{j+1}$ of W_j and W_{j+1} from the dictionaries D_k and D_{k+1}. We calculate the conditional probabilities of symbols in Step 27 by using Equation 1. We calculate the overall probability of S_i using the Equation 2:

Algorithm 2: Markovian Based Algorithm (ID, S, k, N)

Data: IncidentID (ID= $ID_1....ID_n$) and Sequence of activities (S= $S_1...S_n$) from table.

Result: Top N Anomalous IncidentID.

1 create a empty dictionary D_k, D_{k+1}, D'

2 create an arrayList anomalousIncidents

3 **foreach** IncidentID ID_i in ID **do**

4   S_i = get the sequence corresponding to ID_i

5   set noOfSubsequences = S_i.length - k + 1

6   **foreach** j = 1 to noOfSubsequences **do**

7   read the subsequence (W_j) of length = k starting from j^{th} position in S_i

8   read the subsequence (W_{j+1}) of length = $k+1$ starting from j^{th} position in S_i

9   **if** W_j is not present in D_k **then**

10   add (W_j, 1) as (key, value) pair in D_k

11   **else**

12   add (W_j, value + 1) in D_k

13   **if** W_{j+1} is not present in D_{k+1} **then**

14   add (W_{j+1}, 1) as (key, value) pair in D_{k+1}

15   **else**

16   add (W_{j+1}, value + 1) in D_{k+1}

17 **foreach** IncidentID ID_i in ID **do**

18   S_i = get the sequence corresponding to ID_i

19   set noOfSubsequences = S_i.length - k + 1

20   set anomalyScore = 0.0, prob = 0

21   **foreach** j = 1 to noOfSubsequences - 1 **do**

22   read the subsequence (W_j) of length = k starting from j^{th} position in S_i

23   read the subsequence (W_{j+1}) of length= $k+1$ starting from j^{th} position in S_i

24   get the (key_j, $value_j$) pair from D_k corresponding to key = W_j

25   get the (key_{j+1}, $value_{j+1}$) pair from D_{k+1} corresponding to key = W_{j+1}

26   r = ($value_j$) / ($value_{j+1}$);

27   prob = prob + log (r);

28   prob = prob / noOfSubsequences

29   TestSequenceProbability = e^{prob}

30   anomalyScore = 1 / TestSequenceProbabity;

31   add ID_i, anomalyScore into D'

32 sort D' according to decreasing anomalyScore

33 add top N IncidentID into anomalousIncidents

34 return anomalousIncidents

$$P(S_i) = \prod_{i=1}^{l} P(s_i | s_1 s_2 ... s_{i-1}) \qquad (2)$$

where l is the length of the sequence S_i and s_i is the symbol occurring at position i in S_i [8]. For simplification, we take log on both the sides in Equation 2, the modified equation is used in the Step 28. We normalize the probability in Step 29 to take into account the varying length of sequences. We calculate anomaly score for test sequence S_i as the inverse of the probability of S_i in Step 31. Less probability of the test sequence means more anomaly score. We store the IncidentID and its corresponding anomalyScore in the dictionary D'. Then, we sort the IncidentID's in decreasing order of anomalyScore and return top N IncidentID's.

Table 1. Name, Type and Description of Some of Attributes in Merged table of Interaction_detail and Incident_detail

Attribute Name	Attribute Type	Description
Incident_CIType(Aff)	Nominal	There are 13 distinct types of CIs. Example: software, storage, database, hardware, application.
Incident_CISubType(Aff)	Nominal	There are 64 CI Sub-types. Example : web based, client based, server based, SAP.
Incident_Priority	Nominal	There are 5 categories of priority i.e {1, 2, 3, 4, 5}.
Incident_Category	Nominal	There are 4 Incident Category i.e {Incident, Request For Information, Complaint, Request For Change}
Incident_OpenTime	Date Format	The Open Time of Incident is in 'yyyy-MM-dd HH:mm:SS' format.We convert into timestamp in 'hours.
Incident_HandleTime	Date Format	The Handle Time of Incident is in 'yyyy-MM-dd HH:mm:SS format.We convert into timestampin 'hours.
Interaction_CIType(Cby)	Nominal	There are 13 distinct types of caused by CIs. Example: software, storage, database, hardware, application.
Interaction_CISubType(Aff)	Nominal	There are 64 CI Sub-types. Example: web based, client based, server based, SAP.
Closure Code	Nominal	There 15 distinct types of Closure Code. Example: Unknown, Operator Error, Enquiry, Hardware, Software.
Anomalous	Nominal	{Yes,No}.

5.3   Data Pre-processing

We receive top N anomalous IncidentID's as output from algorithms described in Section 5.2. Our aim is to identify root causes of anomalous incidents. We apply data mining techniques (machine learning) such as decision tree classifier to extract rules describing anomalous behaviour. Input to these techniques requires data to be in a particular format and of high quality. Hence, we apply data pre-processing techniques to bring the data in the required format and also to improve the quality. The data pre-processing helps in improving the accuracy and efficiency of the subsequent mining processes. Data goes through series of steps during pre-processing phase: integration, cleaning and transformation etc.

First, we join two tables 'Interaction_detail' and 'Incident_detail' (refer to Phase 4 of architecture diagram in Figure 1). We use attribute 'RelatedIncident' from Interaction_detail table as foreign key and 'IncidentID' from Incident_detail as primary key to perform the join. We give new name 'Interaction_Incident' to the merged table. The merged table contains all the information for the issues that could not be resolved on first call. We create two copies of table Interaction_Incident: Interaction_Incident_Markovian and Interaction_Incident_Windows. We add new attribute 'Anomalous' to the newly created tables. We make the value of attribute 'Anomalous' as 'Yes' for all the top N anomalous IncidentID's and 'No' for rest of the records. Table 1 represents name, type and description of some of attributes obtained after merging Interaction_detail and Incident_detail. The anomalous IncidentID's for table Interaction_Incident_Windows are obtained from the outcome of Algorithm 1. The anomalous IncidentID's for table Interaction_Incident_Markovian are obtained from the outcome of Algorithm 2. We use J48 algorithm for classification using decision tree in Weka. The J48 algorithm handles missing values itself by replacing them with the most frequent observed non-missing values.

Next we transform open time, close time for Interaction and open time, reopen time, resolved time, closed time for Incident given in datetime format. We covert the datetime format that is 'yyyy-MM-dd HH:mm:SS' into timestamp in 'hours' to be useful for classification. For this, we take reference datetime as '1970-01-01 17:13:01'. Handle time for both Interaction and Incident is given in seconds in comma separated format. We transform it by removing comma because the J48 algorithm takes input in CSV or Attribute-Relation File Format (ARFF) format. The pre-processed data serves as input to the classification and visualization techniques.

5.4   Classification

Data mining technique such as decision tree offer a semi automated approach to identify root causes of anomalous incidents. Choosing a data mining analysis tool to execute decision tree algorithm can be a challenge. Popular open source data mining packages include Weka, R, Tanagra, Yet Another Learning Environment (YALE), and Konstanz Information Miner (KNIME)[7]. We choose Weka as it

---

[7] http://www.sciencedirect.com/science/article/pii/S0272271207001114

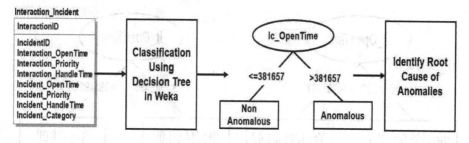

Fig. 4. Flow of classification using Weka

provides a flexible interface which is easy to use. Weka is open source software issued under the GNU General Public License. Weka is a collection of machine learning algorithms for data mining tasks like data pre-processing, classification, clustering, association rules, visualization, etc.

Figure 4 depicts overall flow of classification using decision tree in Weka. The pre-processed data which we obtain after data pre-processing in Section 5.3 serves as input to Weka. The input to Weka is normally in CSV or ARFF format. We use Attribute Selector to select attributes in Weka. Attribute selection involves searching through all possible combination of attributes in the data to find which subset attributes works best for prediction. We perform classification using decision tree algorithm in Weka. Decision tree offers many benefits: easy to understand by user, handles variety of input data such as nominal and numeric and handles missing values in dataset. We apply $J48$ algorithm for decision tree based classification on data. The $J48$ tree classifier is the $C4.5$ implementation available in Weka. The $J48$ builds decision tree from a set of labeled training data using the concept of information entropy. We change the default parameters in $J48$ algorithm like binarySplits, ConfidenceFactor, minNumObj, etc. But, there is no improvement observed in the results. Result is displayed in classifier output window. To view tree in graphical format click on 'visualize tree' option in pop menu.

We consider combination of attributes or parameters as root causes of anomalous incidents which occur on the path from the root to leaf showing anomalous as Yes. Figure 5 and 6 shows the fragments of the decision tree extracted from Weka (due to limited space it is not possible to display the entire tree). To represent figures more clearly, Incident is written as Ic and Interaction is written as Ir. We create decision tree in Figure 5a and 5b by applying $J48$ algorithm on records from Interaction_Incident_Markovian in CSV format. We observe that there are 240 anomalous incidents whose incident open time is greater than 381657 (hrs). Figure 5b shows that there are 38 anomalous incidents whose interaction open time is greater than 383499 (hrs). Decision tree in Figure 6 is for results from Interaction_Incident_Windows in CSV format. Figure 6 depicts there are 67 anomalous incidents whose interaction open time is greater than 384871 (hrs) and incident open time is greater than 384822 (hrs). We obtain

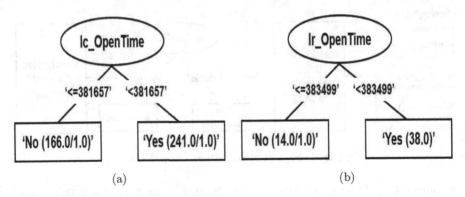

Fig. 5. Fragments of decision tree using Weka based on anomalous incidents received from Markovian based technique

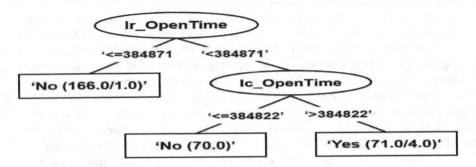

Fig. 6. Fragment of decision tree using Weka based on anomalous incidents received from Window based technique

attributes Incident_OpenTime, Interaction_OpenTime, Incident_Priority and Incident_Category on the path which leads to anomalous incident leaf nodes. And, there is a path consisting of only Incident_OpenTime which classifies 240 incidents as anomalous. Therefore, open time of incidents alone or combination of open time of interaction, open time of incident, priority of incident and category of incident are causes behind anomalous incidents.

5.5 Visualization

Visualization techniques are used to facilitate user interaction with data. User analyze data by carefully examining it and using different tools on it. Visualization techniques help to identify usual trends and anomalies which are present in data. To achieve this, We use the Tibco's Spotfire platform. Tibco Spotfire is an analytics software that helps quickly uncover sights for better decision making. It is used to detect patterns and correlations present in the data that were hidden in our previous approach using Decision tree. Among many features provided by Spotfire, we use Parallel Coordinate Plot for visualization.

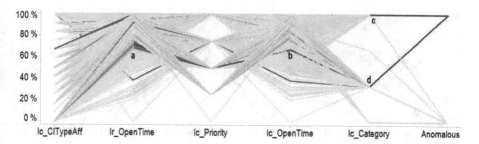

Fig. 7. Parallel coordinate plot depicting the behaviour of incident CI type affected, interaction open time, incident priority, incident open time and incident category for anomalous and non-anomalous incidents from Markovian technique

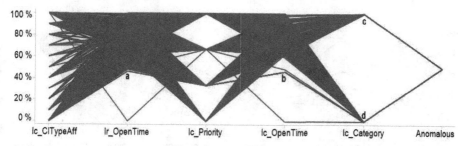

Fig. 8. Parallel coordinate plot depicting the behaviour of incident CI type affected, interaction open time, incident priority, incident open time and incident category for anomalous incidents from Markovian technique

Parallel Coordinate Plot maps each row in the data table as a line. Each attribute of a row is represented by a point on the line. The values in Parallel Plot are normalized. It means lowest value for an attribute in the column is 0% of entire data values in that column while the highest value is 100% unlike the line graphs. Therefore, we cannot compare the values in one column with the values in other column. The data fed into parallel plot is the integrated data which we obtain after the pre-processing phase described in Section 5.3. We create parallel plot by following four steps.

1. Load the data into the Tibco Spotfire. Data can be of type Spotfire Binary Data Format (SBDF), TXT, XLSX, CSV. etc.
2. Choose Parallel Coordinate Plot from Insert tab.
3. Select attributes from column option of the properties section. The selected attributes will be displayed on X-axis of plot.
4. Color the lines or profile of each data row depending on attribute value.

We create plots by taking into account different combination of attributes along with the attribute Anomalous (Yes/No). Figure 7 represents patterns of the attributes: Ic_CITypeAff, Ir_OpenTime, Ic_Priority, Ic_OpenTime and

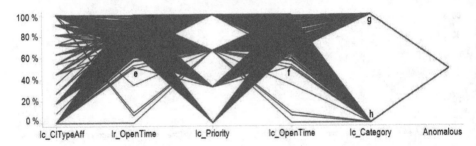

Fig. 9. Parallel coordinate plot depicting the behaviour of incident CI type affected, interaction open time, incident priority, incident open time and incident category for anomalous incidents from Window based technique

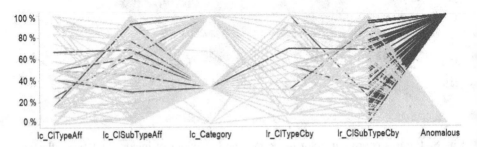

Fig. 10. Parallel coordinate plot depicting the behaviour of incident CI type affected, incident CI subtype affected, incident category, interaction CI type affected and interaction CI subtype affected for anomalous and non-anomalous incidents from Markovian technique

Ic_Category for the complete dataset on parallel plot. The dataset consists of records from table Interaction_Incident_Markovian created in Pre-processing phase in CSV format. Due to scarcity of space, Incident is written as Ic and Interaction is written as Ir. We show anomalous incidents with red color and non-anomalous with yellow color. We consider only anomalous incidents in Figure 8 for better clarity. The attributes Ic_CITypeAff and Ic_Priority individually do not show any useful information regarding anomalous behavior of incidents. Anomalous Incidents are falling in all the Ic_CITypeAffs and Ic_Prioritys, therefore they alone cannot be cause of anomalies. Majority of anomalous incidents are lying above 'a for the attribute Ir_OpenTime. According to it for all anomalous incidents, Ir_OpenTime is above 376305 (hrs) and majority of them have Ir_OpenTime above 381658 (hrs). Point 'b shows that majority of anomalous incidents Ic_OpenTime above 381658 (hrs). Points 'c and 'd depicts that out of 4 Ic_Category : Incident, Complaint, Request for Information and Request for Change, anomalous incidents fall into only 2 categories: Incident and Request for Information. Figure 9 shows parallel plot for the dataset obtained from table Interaction_Incident_Windows. We consider only records which have

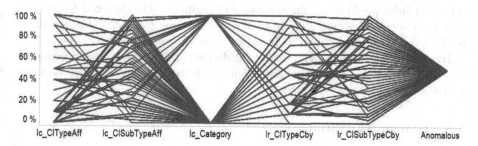

Fig. 11. Parallel coordinate plot depicting the behaviour of incident CI type affected, incident CI subtype affected, incident category, interaction CI type affected and interaction CI subtype affected for anomalous incidents from Markovian technique

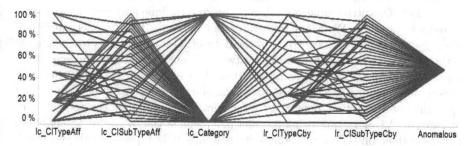

Fig. 12. Parallel coordinate plot depicting the behaviour of incident CI type affected, incident CI subtype affected, incident category, interaction CI type affected and interaction CI subtype affected for anomalous incidents from Window based technique

value of attribute Anomalous as 'YES'. Figure 9 shows that majority of anomalous incidents are lying above 'e for the attribute Ir_OpenTime. According to it for all anomalous incidents, Ir_OpenTime is above 37082 (hrs) and majority of them have it above 381512 (hrs). Point 'f shows that majority of anomalous incidents Ic_OpenTime above 381512 (hrs). Points 'g and 'h depicts that out of 4 Ic_Category's anomalous incidents fall into only 2 categories: Incident and Request for Information.

We choose attributes Ic_CITypeAff, Ic_CISubTypeAff, Ic_Category, Ir_CITypeAff and Ir_CISubTypeAff for Figure 10, 11 and 12. Figure 10 represents parallel plot for complete dataset with anomalous incidents obtained from Markovian technique. Figure 11 and 12 represents parallel plot only for anomalous incidents obtained from Markovian and Window based technique respectively. The selected attributes do not show any useful information regarding anomalous behavior of incidents. Anomalous Incidents are falling in all values for Ic_CITypeAff, Ic_CISubTypeAff, Ir_CITypeAff and Ir_CISubTypeAff. Therefore, they cannot be the cause of anomalies. It concludes that Affected Configuration Item's (CI) type and subtype do not influence cause of anomaly. The Affected Configuration Item is the CI where a disruption of ICT service is noticed. The attribute Ic_Category is showing the same behavior as in Figure 7, 8 and 9.

By comparing the parallel plots for Markovian and Window based technique, it is evident that anomalous incidents from both the techniques follow the same patterns.

5.6   Triangulation Study

In this Subsection, we present our approach on validating the uncovered root cause. In our experiments, we use a publicly available dataset and we do not have facts to validate the root-cause. The real source of the problem is confidential and not known to us or publicly available. We apply data triangulation technique consisting of gathering evidences from multiple sources to validate the root cause[8]. Data triangulation is a well-known technique and we believe is well-suited for our study. We define two evaluators: outcome from parallel coordinate plots and output of decision trees on the dataset. Our objective is to investigate if the findings and indicators from the two different evaluators converge to the same conclusion. Decision tree results described in Section 5.4 show that root causes of anomalous incidents are open time of incidents alone or combination of open time of interaction, open time of incident, priority of incident and category of incident. Visualization using parallel plot depicts that cause of anomalies is not dependent on Affected Configuration Item's (CI) type and subtype which confirms with decision tree results. The parallel plot results also show that root cause of anomalies is dependent on open time of incident, open time of interaction and category of incident (refer to Section 5.5). The experimental results from the decision tree are in agreement with the parallel plot results, thereby validating our approach.

6   Conclusion

We present a novel approach for identification of anomalous traces and executions from event-logs generated by Process Aware Information Systems (PAIS) and a new technique for Root Cause Analysis (RCA) of anomalous traces. The key components of the proposed framework are: anomaly detection from sequential dataset using Window-based and Markovian-based technique, extraction of rules and characteristics of anomalous traces using decision-tree classifiers and application of parallel co-ordinate plots to visualize distinctions between anomalous and normal traces. We conduct a series of experiments on real-world dataset and conduct a triangulation study to demonstrate that the proposed approach is effective. Experimental results reveal agreement in output from Window-based and Markovian technique increasing the confidence in the classification result. We observe that data pre-processing and transformation is needed and impacts the outcome of parallel coordinate plot and decision tree classifier.

---

[8] http://en.wikipedia.org/wiki/Triangulation_(social_science)

References

[1] Van der Aalst, W.: Process mining: discovery, conformance and enhancement of business processes (2011)

[2] Bezerra, F., Wainer, J.: Anomaly detection algorithms in logs of process aware systems. In: Proceedings of the 2008 ACM Symposium on Applied Computing, pp. 951–952. ACM (2008)

[3] Bezerra, F., Wainer, J.: Fraud detection in process aware systems. International Journal of Business Process Integration and Management 5(2), 121–129 (2011)

[4] Bezerra, F., Wainer, J.: A dynamic threshold algorithm for anomaly detection in logs of process aware systems. Journal of Information and Data Management 3(3), 316 (2012)

[5] Bezerra, F., Wainer, J.: Algorithms for anomaly detection of traces in logs of process aware information systems. Information Systems 38(1), 33–44 (2013)

[6] Bezerra, F., Wainer, J., van der Aalst, W.M.P.: Anomaly detection using process mining. In: Halpin, T., Krogstie, J., Nurcan, S., Proper, E., Schmidt, R., Soffer, P., Ukor, R. (eds.) Enterprise, Business-Process and Information Systems Modeling. LNBIP, vol. 29, pp. 149–161. Springer, Heidelberg (2009)

[7] Calderón-Ruiz, G., Sepúlveda, M.: Automatic discovery of failures in business processes using process mining techniques

[8] Chandola, V., Banerjee, A., Kumar, V.: Anomaly detection for discrete sequences: A survey. IEEE Transactions on Knowledge and Data Engineering 24(5), 823–839 (2012)

[9] Forrest, S., Hofmeyr, S., Somayaji, A., Longstaff, T.: A sense of self for unix processes. In: Proceedings of the 1996 IEEE Symposium on Security and Privacy, pp. 120–128 (May 1996)

[10] Heravizadeh, M., Mendling, J., Rosemann, M.: Root cause analysis in business processes (2008)

[11] Hofmeyr, S.A., Forrest, S., Somayaji, A.: Intrusion detection using sequences of system calls. Journal of computer security 6(3), 151–180 (1998)

[12] Ron, D., Singer, Y., Tishby, N.: The power of amnesia: Learning probabilistic automata with variable memory length. Machine Learning 25(2-3), 117–149 (1996)

[13] Suriadi, S., Ouyang, C., van der Aalst, W.M., ter Hofstede, A.H.: Root cause analysis with enriched process logs. In: Business Process Management Workshops, pp. 174–186 (2013)

[14] Vasilyev, E., Ferreira, D.R., Iijima, J.: Using inductive reasoning to find the cause of process delays. In: 2013 IEEE 15th Conference on Business Informatics (CBI), pp. 242–249. IEEE (2013)

[15] Wainer, J., Kim, K.-H., Ellis, C.A.: A workflow mining method through model rewriting. In: Fukś, H., Lukosch, S., Salgado, A.C. (eds.) CRIWG 2005. LNCS, vol. 3706, pp. 184–191. Springer, Heidelberg (2005)

Modeling Personalized Recommendations
of Unvisited Tourist Places Using Genetic Algorithms

Sunita Tiwari[1] and Saroj Kaushik[2]

[1] School of IT, IIT Delhi, New Delhi, India 110016
sunita@cse.iitd.ac.in
[2] Dept. of Computer Science and Engg., IIT Delhi, New Delhi, India 110016
saroj@cse.iitd.ac.in

Abstract. Immense amount of data containing information about preferences of users can be shared with the help of WWW and mobile devices. The pervasiveness of location acquisition technologies like Global Positioning System (GPS) has enabled the convenient logging of movement histories of users. GPS logs are good source to extract information about user's preferences and interests. In this paper, we first aim to discover and learn individual user's preferences for various locations they have visited in the past by analyzing and mining the user's GPS logs. We have used the GPS trajectory dataset of 178 users collected by Microsoft Research Asia's GeoLife project collected in a period of over four years. These preferences are further used to predict individual's interest in an unvisited location. We have proposed a novel approach based on Genetic Algorithm (GA) to model the interest of user for unvisited location. The two approaches have been implemented using Java and MATLAB and the results are compared for evaluation. The recommendation results of proposed approach are comparable with matrix factorization based approach and shows improvement of 4.1 % (approx.) on average root mean squared error (RMSE).

Keywords: Location Recommender Systems; GPS Log Mining; Soft Computing; Genetic Algorithms; User Preference Discovery.

1   Introduction

Ease of location acquisition technologies like Global Positioning Devices (GPS) has generated a lot of interest amongst researchers to mine and learn patterns of human behavior from spatio temporal data and their movement histories (GPS trajectories). GPS data contains rich information about human activities and preferences. People share their GPS logs and other personal information on the internet. This information can be used to infer how people move around and extract their context, habits and likings. Analysis of this inferred information is an important research problem which can find applications in recommending interesting locations to the tourists, traffic planning, itinerary planning, recommending advertisement hoardings sites etc.

One can extract popular and significant locations (places) by analyzing the GPS traces of multiple users [5], [17]. Incorporating semantic information about the

W. Chu et al. (Eds.): DNIS 2015, LNCS 8999, pp. 264–276, 2015.

locations may further improve the ranking of popular locations [11]. Here popular locations are places which are frequently visited and liked by a large number of people. Examples of such locations are historical monuments, temples, parks, restaurants, shopping malls etc. There has been lot of prior work on mining popular (interesting) locations in a geo spatial region [2], [5], [17]. In most of these approaches, a location which is visited by many people is considered as an interesting location. But in contradiction, locations frequently visited by several users may not be popular or attractive such as a traffic signal, a busy street, offices etc. The authors proposed a semantic annotation based approach to address this issue in [11].

As explicit user ratings are not easily available for all the domains, the implicit elicitation of user preferences is essential for personalized applications. For such domain, user's preferences are learnt from user's behaviour pattern. For recommending popular location to a user we need to infer user's interest. Our work in this paper is an endeavour to learn and infer the interest of individual users in already visited locations by mining individual user's location histories (GPS logs). Application of such information can be found in understanding the correlation between users and locations, travel recommendation, learning travel and behaviour pattern of people, traffic planning, mobile tourist guide etc.

Further, we have proposed a Genetic algorithm (GA) based approach for recommendation in order to evolve the user's interest to an unvisited location. GA being an optimization technique has been used for finding the best and optimized recommendation. The contribution of this paper is twofold:

- A novel elicitation approach for mining user's preference using his/her GPS logs.
- GA based approach to evolve best locations for recommendation.

The rest of the paper is organized as follows. Section 2, provides a brief description of related work, section 3, discusses the overall system design. Section 4, illustrates experimental results and discussions. Finally, section 5, concludes the paper and presents future directions.

2   Related Work

There has been a lot of prior work on mining, analyzing behavior history using GPS data and analyzing multiple users' information to learn patterns. Various approaches are also found for location recommendation and some of them are reviewed below:

Two major contributions by mining GPS traces of multiple users for interesting location recommendations are made in [15], [17]. These contributions include 1) providing the top interesting locations recommendations and travel sequences in a given geospatial region. Hierarchical graph based approach is used to find interesting locations in a geospatial region. 2) A personalized recommender system that provides an individual with locations matching her travel preferences. Exploiting correlation between different locations, item based collaborative filtering model and users' past location histories generate recommendations. For evaluation of the system, real world GPS trace collected for 107 users over a period of one year has been used.

A framework for friend and a location recommendation based on individual user's GPS traces have been proposed in [12]. The friend recommendation is made for social networks. Past location histories of individual users are used to measure similarity among those users. A hierarchical graph based similarity measure is proposed. Using these past location histories, the estimate of individual's interest in unvisited places is predicted. GPS data collected by 75 subjects over a period of 1 year in the real world is used for the evaluation of the proposed framework.

CityVoyager [10] is a recommender system which recommends shops to users based on their individual preferences and needs. The individual preferences are estimated by analyzing past location histories of users acquired using GPS enabled devices. Two shops namely, 'A' and 'B' are considered similar if there is an observed tendency that user who frequently visit shop 'A' also frequently visit shop 'B'. Location correlation concept is used for making recommendations. A similar work based on location correlation can be found in [16].

User preferences changes drastically with the changes in time and location. Therefore the personal preferences should be clubbed with the location information in order to improve accuracy of recommendation. Understanding the user's behaviour using GPS history and eliciting accurate user preferences is a non trivial task. Various data mining and machine learning techniques can be used for inferring the user preferences automatically. Zenebe et al. [13] proposed an approach to predict user preferences with uncertainty. Also they have considered visualization of item features, user feedback for discovered preferences and improve the interpretation of the discovered knowledge. M. F. Mokbel et al [8] user preferences and context are taken into account to answer user queries. They mostly have adapted existing commercially successful recommendation techniques to consider the spatial aspects of users and/or items when making its decisions. Online location based social networks have been exploited for inferring the implicit user ratings in [4].The frequency of check in to a spot is considered as user rating for that location. Use of matrix factorization technique for spot recommendation has been proposed in this work.

Other works include mobile tourist guide systems [1], [3], [7], [9], which recommend locations and/or provide navigation information based on a user's current location. Other work includes detecting user behaviour based on individual location history represented by GPS trajectories [14], [15].

3   Overall System Design

In this section we will discuss the overall design of the proposed recommendation system (shown in figure 1) and working of the various modules. The proposed system consists of five modules namely, Stay Point Generation, Popular Location Extraction, Implicit Rating Elicitation, Prediction and Recommendation. These modules are described as follows:

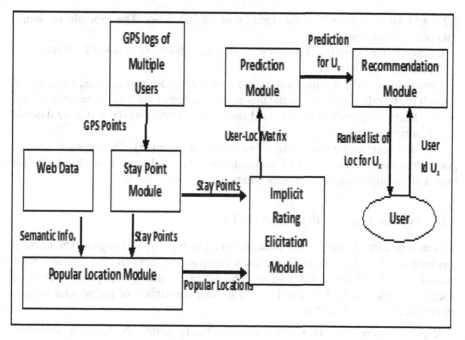

Fig. 1. Overall System Design

3.1   Stay Point Generation (SPG)

The stay point generation module computes stay points for every user. A geographical location where a user spent some considerable amount of time (greater than a specific threshold) is called a stay point. Examples of stay points include traffic signals, historical locations, restaurants, homes, offices etc. Substantial stay of a user at a particular point may be because of the following two situations [14].

1. The user remains stationary for a substantial amount (threshold τ) of time. This may happen due to the loss of GPS signal when a user enters inside a building (GPS devices loose satellite signal indoors). In this case, no GPS points are generated for a considerable amount of time.
2. User wanders around within some spatial region (within some considerable distance, threshold δ) for a period. In this period, the device may generate several GPS points in a spatial region and stay point corresponding to that sub trajectory, can be computed by taking the average of these GPS points or some other method such as centroid.

We have used the GPS trajectory dataset of 178 users collected by Microsoft Research Asia's GeoLife project collected in a period of over four years [18]. These trajectories were recorded by different GPS loggers and GPS phones. Most of the trajectories are logged in a dense representation, e.g. every 1~5 seconds or every 5~10 meters per point. This dataset contains 17,621 trajectories with a total distance of

1,251,654 kilometres and a total duration of 48,203 hours. The example of sample data point is shown as follows.

Example: "39.906631, 116.385564, 0, 492, 40097.5864583333, 2009-10-11, 14:04:30".

Here fields 1 and 2 are the latitude and longitude in decimal degrees. Field 3 is all set to 0 for this dataset. Field 4 contains altitude in feet; field 5 is the number of days (with a fractional part) that have passed since 12/30/1899. Fields 6 and 7 are date and time represented as a string.

The stay point computation algorithm is described in [11]. The inputs to the stay point algorithm are GPS logs of users, distance threshold and time threshold. The output of this algorithm is set of stay points.

3.2   Popular Location Extraction (PLE)

This module extracts the popular locations in a given geo spatial region. The inputs to this module are stay points of all the users, computed by SPG and the semantic information from the different web sources. The detailed working of popular location extraction module can be found in [11]. The brief description of the popular location extraction procedure is as follows:

3. Cluster the stay points into locations (places) using DBSCAN clustering algorithm.
4. Find the centre of each cluster and its corresponding address using reverse geocoding process.
5. Attach semantic annotation/tags to each popular location by using the ontology based focused crawler.
6. Using semantic tags classify the locations into popular and unpopular locations.

For the extracted set of popular locations we further aim to find implicit user rating for locations which has been visited by these users and predict the interest of users to yet unvisited locations.

3.3   Implicit Rating Elicitation Module

In scenario when user's explicit ratings are not available, the recommendation generation is a harder task as compared to explicit feedback/rating based systems. To address this problem implicit ratings are learnt using user behaviour pattern such as user movement histories. We have proposed an approach for mining implicit user rating from user GPS logs. Inputs to this module are 1) stay points of individual user generated by SPG module and 2) popular locations generated by PLE module.

For implicit rating of a user U, the number of check in (frequency of visits) to every visited cluster (popular locations) L are counted. For this purpose we have classified the stay points of individual user into one of the clusters formed in PLE module. Thus we obtain the frequency for each user to all the visited locations. The frequency data shows that users have visited very few locations frequently.

Once we have obtained the frequency for each user U to every location L, we normalize these values in range [0, 1]. To generate a model from the extracted information, we represent data in the form of user location matrix as depicted in figure 2. Each row of the given matrix corresponds to a user, and each column corresponds to a location. These entries of the matrix represent a user's interest in a location, with high values representing high and low values representing low interest. The entry Pij represents the interest of user i in location j. The cells containing '?' in the matrix are not yet visited locations. Now we will use this matrix to predict values for unvisited locations.

3.4   Prediction Module

This module predicts the missing values in user location matrix. Example user location matrix is shown in figure 3 which shows a part of user location matrix from our experiments. This matrix contains the implicit rating of users for visited popular locations. The null value in i^{th} row and j^{th} column represents that user i has not yet visited location j.

	L_1		L_j		L_m
U_1	$P_{1,1}$		?		$P_{1,m}$
U_i	$P_{i,1}$		?		?
U_n	$P_{n,1}$		$P_{n,j}$		$P_{n,m}$

Fig. 2. User Location Matrix

Genetic Algorithm (GA) is a powerful tool for solving search and optimization problems. The basic concept behind GA is that the strong have a tendency to adapt and survive whereas the inferior tend to die out (survival of fittest). In order to implement GA we need to define encoding scheme, fitness function, selection, crossover and mutation operators. The architecture of our GA based approach is discussed as follows.

Chromosome Encoding: The GA chromosome structures consist of a matrix M of size |u| ×|l|, where |u| is number of user and |l| is a number of popular locations. Every entry $M_{i,j}$ in matrix(chromosome) consists of values in range [0, 1] which represents the interest of the user ui in location lj. shows an example of chromosomes used.

Initial Population: Initial population consists of n matrices (similar to one shown in figure 4) of size |u| ×|l|, containing random values in the range [0, 1].

	L_1	L_2	L_3	L_4
U_1	1	0.6	?	0.2
U_2	0.8	?	?	0.2
U_3	0.2	0.2	?	1
U_4	0.2	?	?	0.8
U_5	?	0.2	1	0.8

Fig. 3. User Location Matrix M1

Fitness Function: We have used root mean squared error (RMSE) as a fitness function. The fitness value for the kth candidate solution is computed as

$$f^k = \sum_{i,j} (P_{i,j} - P_{i,j}^k)^2$$

(1)

	L_1	L_2	L_3	L_4
U_1	0.97	0.61	0.418	0.197
U_2	0.788	0.24	0.199	0.197
U_3	0.221	0.197	0.55	0.981
U_4	0.2	0.88	0.44	0.811
U_5	0.12	0.221	1	0.788

Fig. 4. Example Chromosome

Here $P_{i,j}$ represents the *actual rating* in a user location matrix's (similar to matrix M1 as shown in figure 3) i^{th} row and j^{th} column and $P_{i,j}^k$ represents the corresponding entry in the kth *candidate solution*. Only entries, containing actual ratings in user location matrix (as well as in corresponding candidate solutions) are considered in fitness evaluation so '?' entries are ignored while evaluating fitness of the chromosome. For example, consider the candidate solution given in figure 4, the fitness value is computed as f=0.001662. Our objective here is to minimize the value of fitness function f.

Selection: This selection is done depending on the relative fitness of the individuals so that best ones (with minimum fitness values) are chosen for reproduction than poor ones.

Crossover: Next, after selecting two parents, offspring are bred by the selected individuals. **Crossover** operator is used to generate offspring. For generating new chromosomes, our algorithm used **block uniform crossover** (exchange occurs inside a rectangular block whose size and location are determined randomly). In the block uniform crossover, each block (of size i x j, where i, j are chose randomly) of parent is interchanged randomly with the corresponding block of second parent based on pre-assigned probability. The fitness of the new chromosomes is evaluated using eq 1. Next, the individuals from the old population are killed and replaced by the new ones.

	L_1	L_2	L_3	L_4
U_1	0.999	0.599	0.436	0.198
U_2	0.797	0.48	0.394	0.199
U_3	0.202	0.187	1.000	0.989
U_4	0.2	0.17	0.918	0.786
U_5	0.272	0.214	0.978	0.824

Fig. 5. Optimal solution computed for user location matrix M1

Convergence Criteria: The algorithm is being stopped when the population converges toward the optimal solution. A solution is considered optimal if the overall difference between already known values of original matrix (say M1given in figure 3) and corresponding values in solution matrix (shown in figure 5) are minimized. Consider the solution given in figure 6 for the example user location matrix M1 shown in figure 3. Here the difference between the known values of matrix M1 at example location (0,0) and corresponding location in candidate chromosome (shown in figure 5) is (1-.999=.001). Similarly the difference at all the already known (i, j)th entries is computed and if the overall difference is minimized then the solution is considered as optimal solution.

3.5   Recommendation Module

Input to the recommendation module is the user id (say U_x) and output is the ranked list of top locations for the individual user Ux. Location recommendation module ranks the list of location for user U_x based on the values predicted by prediction module for U_x and recommends the top k locations to that particular user. This recommendation is personalized for individual user as location histories of U_x is considered filling user location matrix.

4   Experimental Results and Discussions

The GPS trajectories generated by multiple users are mined to extract interesting locations in a given geospatial region. Quality of results is further refined by annotating semantics to the places and comparison of results can be found in [11].

In this paper, we predicted the user's interest in an unvisited location using genetic algorithm and used these results to recommend locations to the users. For the evaluation of the proposed approach, we have implemented the proposed GA based approach and the matrix factorization (MF) technique given in [4]. Both the results are compared and findings of the comparison are discussed below (refer figure 6 and table 2). The most important evaluation metrics for measuring the predictive accuracy of a recommender system are mean absolute error (MAE), mean squared error (MSE), root mean squared error (RMSE) and normalized mean absolute error (NMAE). To measure the accuracy of prediction of our system, we have used Root Mean Squared Error (RMSE) metric which is commonly used in the recommender systems evaluation.

For the illustration, consider the user location matrix M1 given in figure 3. The solution for this matrix using GA and matrix factorization technique [4] is shown in figure 6. The values predicted by both the approaches clearly show that the results are comparable. The RMSE value of GA based solution of M1 is 0.001662 and RMSE of Matrix Factorization based solution of M1 is 0.001762. An improvement of 5.68% is shown for this example.

We carried out experiment to evaluate the effectiveness of our approach and its ability to evolve optimal/near optimal predictions (ratings), The dataset of 178 users and 103 popular locations (which was the output of popular location extraction technique discussed in [11]) was considered. We divided the dataset into five test sets, each set containing random data of 15 users for all 103 popular locations.

	L_1	L_2	L_3	L_4	L_1	L_2	L_3	L_4
U_1	0.999	0.599	0.436	0.198	0.998	0.601	0.437	0.197
U_2	0.797	0.48	0.394	0.199	0.786	0.49	0.392	0.199
U_3	0.202	0.187	1.000	0.989	0.203	0.188	1.000	0.999
U_4	0.2	0.17	0.918	0.786	0.2	0.171	0.92	0.788
U_5	0.272	0.214	0.978	0.824	0.278	0.213	0.976	0.82
	a)				b)			

Fig. 6. Solution for example illustration M1 by a) GA based Approach b) Matrix Factorization based Approach

GA based and matrix factorization based approaches were applied on each set for five different runs. In each run, GA was terminated with less RMSE as compared to MF based approach. The RMSE of Dataset1 for each run is shown in table 1 and plot of percentage improvement is shown in figure 7.

Table 1. RMSE for Dataset1 in different runs

Runs	RMSE for GA based Approach	% Improvement over MF based Approach
1	0.13	3.417533432
2	0.131	2.674591382
3	0.132	1.931649331
4	0.129	4.160475483
5	0.128	4.903417533
Average	**0.13**	**3.417533432**

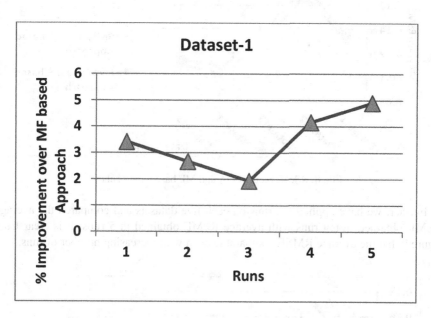

Fig. 7. Plot of percentage improvement of RMSE for Dataset 1

MATLAB was used for implementing both GA and MF based approaches and the performance comparison is shown in figure 8 and table 2. It is observed that an improvement of approx. 4.12% (average RMSE) is achieved by GA based approach over MF based approach.

Table 2. Comparison of Results for GA based and Matrix Factorization Based Approach

Dataset	RMSE for GA based Approach	RMSE for Matrix Factorization based Approach	% Improvement
Dataset1	0.13	0.1346	3.417533432
Dataset2	0.169	0.175	3.428571429
Dataset3	0.13	0.139	6.474820144
Dataset4	0.103	0.109	5.504587156
Dataset5	0.11	0.112	1.785714286

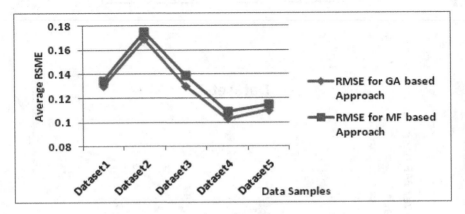

Fig. 8. RSME plots of GA and MF based approaches

Further, we have applied ten runs on each five datasets and compared the average RSME obtained in ten runs with average RSME obtained in 5 runs. It is clear from figure 9 that the average RMSE comparable even with increasing number of runs.

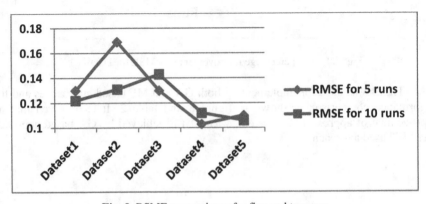

Fig. 9. RSME comparisons for five and ten runs

We have also performed experiments to validate the GA's ability of correct prediction for already visited locations. In initial phase of experiment the preference values (ratings) which are unknown are being evolved. In the second phase of experiment, to test prediction power of GA the evolved values of '?' locations are being used as known ones and the values known previously are assumed to be unknown.

Once we have evolved the optimal user-location matrix, the entry $P_{i,j}$ of user location matrix (representing the interest of user i in already visited locations j) is replaced by '?' and considered the newly evolved entries as known entries. Now we again applied GA approach to evolve these '?'. While predicting the unknown values (which are known previously) we have got the correct results (ratings). The error in predicting already known values is only 3.2% which is fairly acceptable.

5   Conclusion and Future Direction

In this paper, analysis of GPS logs of multiple users is used for location recommendation. The first major contribution of this work is to generate the implicit rating for all visited spot for individual users. For this purpose a user location matrix is generated which contains the implicit rating for the popular locations visited found for all the users. The second contribution of the paper is predicting the interest of user for yet unvisited location using genetic algorithm and finally recommending the top k locations to the user. The results of genetic algorithm based approach are compared with the matrix factorization approach proposed in [4] and an improvement in average RMSE of (approx.) 4.1% is observed for different data samples of dataset tested. The improvement in average RMSE shows that the proposed system has comparable prediction accuracy than the matrix factorization approach.

In future we aim to test and validate our results on more number of users. Also, number of check in alone does not directly reflect the preference of a user to that location and hence we have to make use of other information like temporal information, user demographic information etc. as well. In future we aim to utilize user feedback obtained from blogs and social networks and demographic features to further refine the accuracy of implicitly elicited rating and predicted user ratings. We also aim to improve the computation efficiency by evolving values based on the preferences of similar users.

Acknowledgement. We would like to acknowledge Priti Jagwani, Shivendra Tiwari and Sourabh Tiwari for their support.

References

1. Abowd, G.D.: Cyberguide: a mobile context aware tour guide. Wireless Networks 3(5), 421–433 (1997)
2. Ashbrook, D., Starner, T.: Using GPS to learn significant locations and predict movement across multiple users. Personal and Ubiquitous Computing 7(5), 275–286 (2003)

3. Beeharee, A., Steed, A.: Exploiting real world knowledge in ubiquitous applications. Personal and Ubiquitous Computing 11(6), 429–437 (2006)
4. Berjani, B., Strufe, T.: A Recommendation System for Spots in Location Based Online Social Networks. In: Procedings of SNS 2011, Salzburg, Austria (2011)
5. Khetrapal, S., Chauhan, R., Gupta, S.K., Subramanyam, L.V., Nambiar, U.: Mining GPS Data to Determine Interesting Locations. In: Proceeding of IIWeb 11, Hyderabad, India (2011)
6. Krumm, J.: A Survey of Computational Location Privacy. Personal and Ubiquitous Computing 13(6), 391–399 (2009)
7. Li, X., Mi, Z., Zhang, Z., Wu, J.: A location-aware recommender system for Tourism. In: Proceedings of 2nd IEEE International Conference on Information Science and Engineering (ICISE), Hangzhou, China, pp. 1709–1711 (2010)
8. Mokbel, M.F., Bao, J., Eldawy, A., Levandoski, J., Sarwart, M.: Personalization, Socialization, and Recommendations in Location-based Services 2.0. In: PersDB 2011, Seattle, Washington, USA (2011)
9. Park, M.-H., Hong, J.-H., Cho, S.-B.: Location-Based Recommendation System Using Bayesian User's Preference Model in Mobile Devices. In: Indulska, J., Ma, J., Yang, L.T., Ungerer, T., Cao, J. (eds.) UIC 2007. LNCS, vol. 4611, pp. 1130–1139. Springer, Heidelberg (2007)
10. Takeuchi, Y., Sugimoto, M.: CityVoyager: An outdoor recommendation system based on user location history. In: Ma, J., Jin, H., Yang, L.T., Tsai, J.J.-P. (eds.) UIC 2006. LNCS, vol. 4159, pp. 625–636. Springer, Heidelberg (2006)
11. Tiwari, S., Kaushik, S.: Mining Popular Places in a Geo-spatial Region Based on GPS Data Using Semantic Information. In: Madaan, A., Kikuchi, S., Bhalla, S. (eds.) DNIS 2013. LNCS, vol. 7813, pp. 262–276. Springer, Heidelberg (2013)
12. Zheng, Y., Zhang, L., Ma, Z., Xie, X., Ma, W.-Y.: Recommending Friends and Locations Based on Individual Location History. ACM Transaction on the Web 5(1) (2011)
13. Zenebe, A., Zhou, L., Norcio, A.F.: User preferences discovery using fuzzy models. Fuzzy Sets and Systems 161(23), 3044–3063 (2010)
14. Zheng, Y., Zhou, X.: Computing with Spatial Trajectories. Springer, Heidelberg (2011)
15. Zheng, Y., Xie, X.: Learning travel recommendations from user generated GPS traces. ACM Transaction on Intelligent Systems and Technology (ACM TIST) 2(1), 2–19 (2011)
16. Zheng, Y., Zhang, L., Xie, X., Ma, W.Y.: Mining correlation between locations using human location history. In: Proceedings of the 17th ACM SIGSPATIAL International Conference on Advances in Geographic Information Systems 2009, pp. 472–475 (2009)
17. Zheng, Y., Xie, X., Ma, W.-Y.: Mining Interesting Locations and Travel Sequences from GPS Trajectories. In: proceedings of ACM International World Wide Web Conference (WWW 2009), Madrid, pp. 791–800 (2009)
18. Zheng, Y., Li, Q., Chen, Y., Xie, X., Ma, W.-Y.: Understanding Mobility Based on GPS Data. In: Proceedings of ACM Conference on Ubiquitous Computing (UbiComp 2008), Seoul, pp. 312–321 (2008)

A Decentralised Approach to Computer Aided Teaching via Interactive Documents

Lothar M. Schmitt

The University of Aizu, Aizuwakamatsu 965-8580, Japan
L@LMSchmitt.de

Abstract. We demonstrate how certain interactive features of learning management systems, *i.e.*, quizzes and certain types of exercises including evaluations and feedback, can be incorporated into LaTeX-generated pdf-documents via embedded javascript-code. The embedded computational abilities of the enhanced pdf-documents which are discussed in this work allow for the repeated creation and presentation of exercises (within one session of use of such the document), if the exercises are designed to contain components that are randomly generated or randomly selected from a limited family of predesigned choices stored within the document. This enables the user/student to use the embedded exercises in the described interactive teaching documents for drill purposes. In addition, the use of such documents as extensions of common learning management systems is discussed.

Keywords: interactive documents, learning management systems, mathematics, latex, pdf-forms, javascript.

Introduction

This work discusses the porting of certain aspects and components of interactive teaching systems developed in [22], [12] and [20] into interactive teaching documents in pdf format, in particular, for the teaching of mathematics. The latter teaching documents can be used in a stand-alone fashion, independent of the internet and a central server which hosts a particular teaching system. However, communication with a central host which registers and administers, *e.g.*, work done using such interactive teaching documents as part of a university course can also be achieved (see, *e.g.*, [18] and section 3).

The teaching documents are developed using the LaTeX typesetting environment which is[1], at this point in time, best suited for typesetting mathematical treatise. As output, the typesetting environment produces pdf-files in which interactive elements such as pdf-forms [1, sec. 8.6] are embedded. Among several types of similarly implemented interactions, the documents contain input-windows where the reader is suppose to enter the solution to an exercise, and output-windows where the computational engine within the documents (based

---

[1] In the *opinion* of this author.

W. Chu et al. (Eds.): DNIS 2015, LNCS 8999, pp. 277–288, 2015.

upon embedded javascript-code) returns an evaluation of the reader's input including appropriate comment. The computational engine within the documents can generate and present randomly generated problems repeatedly to the reader within one session of use such that the reader can practice solving a certain type of problem multiple times. In particular, the document presents a new or altered set of problems, every time it is reopened.

The 1990's saw the world-wide explosive expansion of the internet due to the invention of the first generation of web-browsers, *i.e.*, Netscape and Mosaic. As part of this revolution, a multitude of learning management systems (commercial or public-domain) were published. Examples for early-existing and presently-existing learning management system are: Blackboard (commercial [4]), Moodle (open source [16]), and UNEIM² [22],[12],[20]. As most learning management systems, Blackboard and Moodle operate from a host website which can be accessed by students via the internet using a web-browser. In contrast to these systems, UNEIM uses email as means of communication between students and the learning management system.

Today, there are a vast number of learning management systems (both commercial and open source) available on the market [28]. However according to [15], Moodle (1) and Blackboard (3) are still today leading among the most popular learning management systems. We observe that Moodle's built-in default ability to handle and generate quizzes or exercises is described by "calculated, description, essay, matching, embedded answers, multiple choice, short answer, numerical, random short answer matching, true/false" according to [29, sec. 4], compare also [17]. According to [6], Blackboard's ability to incorporate interactive quizzes and exercises into the system covers an almost identical range of possibilities. Let us note that all these types of quizzes and exercises listed in [17] and [6] can be implemented (and partially have been implemented) in UNEIM.

In this work, we intend to outline how quizzes and exercises with random components similar to quizzes and exercises within Moodle, Blackboard, and UNEIM can be ported into interactive teaching pdf-documents which offer the same functionality as the quizzes and exercises within the latter three learning management systems. See section 2.

Finally, let us remark at this point that all figures of this document are collected in an appendix at the end of the paper.

1   The UNEIM/MAT System and Its Evolution

As described in [22] and [20], the UNEIM system is a learning management system which was originally conceived around 1995 to assist the teaching of English Composition to Japanese students, in particular, in the setting of The University of Aizu where the Center for Language Research is fully embedded within the university's UNIX-based [13] computer network. Initially, the intended application for the UNEIM system was to relieve the instructor of a course English

---

² Unix Network English Instruction Manager.

Composition in the field of teaching English as a Foreign Language from repetitive tasks such as correcting basic grammar and vocabulary mistakes and, at the same time, to administer such a course to a larger extent. Input to the UNEIM system are students' essays sent to an instructor's account by email. Output of the UNEIM system are various results of automated scans of the submitted essays ranging from a simple spell-check to pattern analysis of text in order to find typical mistakes by Japanese students such as the use of the phrase "play judo" [26].

Note that the UNEIM system allows for the students multiple and, hopefully, continuously improved submissions of their homework, and thus, provides a greater amount of feedback than what is humanly achievable.

In order to administer a university course, the UNEIM system contains a database related to various aspects of the administration process such as portfolios of students' contributions, combined lists of results or histories of evaluations, and others. Due to its relatively small size, UNEIM's database is implemented as a family of plain files (containing essays, evaluations, flags, and lists) and folders (pertaining to students, sessions, courses, semesters and academic years) which are traversed by certain evaluation and administrative programs in a periodic fashion.

Around the year 2000, this author extended the UNEIM system in such a way that it became suitable to assist in teaching and administering a course in mathematics. This extension, labeled MAT[3], is described in detail in [20]. Initially, the mathematics-related part of the MAT system[4] sends out a tasksheet to the students of a course plus an ASCII-text-only answer-form which the students are supposed to return to the instructor's account via email. The submitted answer-forms are then scanned by evaluation programs, and corrected answers and appropriate comment (at least in regard to the first mistake of a particular student in an exercise) are mailed back to the students to trigger an improved (re-)submission followed by a new evaluation plus further machine feedback.

(1.1) The following types of problems are generally checked by the MAT-system in practice:

(1.1.1) Simple yes/no, or multiple-choice type questions.

(1.1.2) Problems which request the return of a numerical result such as the coefficients of the inverse of a matrix, or the return of a list of successive states of an algorithm.

(1.1.3) Problems which request the return of a formula in simplified ASCII-character format (*e.g.*, similar to Mathematica input-format [30]).

(1.1.4) Problems which request the return of LaTeX code. For example, the proof of $\lim_{n\to\infty} n^{-1} = 0$ is given, including the LaTeX code (devided into fragments representing the lines of the proof), and the student is supposed

---

[3] Machine Assisted Teaching.

[4] The English Composition related portion of UNEIM is incorporated in MAT as well.

to return a small variation of the given LATEX code line-wise which proves $\lim_{n\to\infty} n^{-2} = 0$.

(1.1.5) The carefully crafted lines of a proof of a stated proposition are presented in a permuted order as a numbered list, and the student has to return the proper order of the lines of the proof. This can be varied in that the concatenated lists of lines of two proofs are presented in a permuted order as a numbered list, and the student has to assemble two proofs of two given propositions from that.

Note that the techniques for quizzes and exercises outlined in (1.1.1)-(1.1.5) can, quite obviously, be used to generate quizzes and exercise for other subjects than mathematics.

In [12], the approach listed in (1.1.1)-(1.1.5) above was "randomised," i.e., certain tasks were generated with random coefficients or randomly generated objects, such that every student in a class' exercises administered by the MAT system can be served with a unique set of problems. While this author pursues the teaching philosophy that MAT-administered exercises and quizzes are (only) an opportunity to practice for the final exam which determines the grade, individualised tasks, obviously, provide a strong incentive for students to really work on problems rather than copying solutions from others in order to show participation in the class.

Note that, in principle, problem types (1.1.1)-(1.1.5) could be implemented as Moodle quizzes according to [17], but the randomised versions of quizzes as in [12] cannot. For example, it seems very hard to randomly create a finite state machine which *can* be reduced using the machine reduction algorithm as a problem instance in Moodle.

2   Incorporating Randomised Problems into Interactive Documents

In view of the saturation of society with tablet computers and phablets (*i.e.,* large-screen phones) as well as netbooks and laptops, it seemed desirable a few years ago to try to port part of the UNEIM/MAT framework (including the randomised problem generation) in such a way that the latter devices could be used for independent study[5] of (elementary[6]) mathematics and other subjects, even offline. Furthermore, if such teaching documents are equipped with a free sharing license such as the GNU-GPL [9], then the documents and associated the educational process can propagate through society worldwide via personal

---

[5] This being consistent with the founding motto of The University of Aizu: To Advance Knowledge for Humanity.

[6] In the experience of this author, it is very difficult, if not impossible, to formulate, grade and manage meaningful exercises for higher level mathematics whithin the MAT framework. This holds, in particular, when proving complicated propositions is part of the exercise, since there are usually many ways one can find and write a correct proof of a particular statement.

file sharing and/or posting such a document on various dedicated servers independently from the author's efforts.

As a first project realising this approach to distributed teaching, mainly software engineering work in [11] implants a collection of interactive exercises designed by the author of [11] and this author into a treatise on elementary set theory [21]. In what follows, we shall illustrate some of the types of exercises which were implemented in [11] for elementary set theory [21]:

(2.1) Analogue to section (1.1.1) and the work in [23], [24] and [25], a yes/no type quiz environment was implemented for some of the exercises in [21]. The difference in regard to the work in [23], [24] and [25] is however, that comment can be added to every question which is displayed in LATEX-quality including formulas, after the question has been answered (correctly or not) via pressing either a yes- or a no-button. Compare also [14]. Figure 1 shows an example in this regard[7].

(2.2) Analogue to section (1.1.1), but not static as in section (2.1) or [14], randomly generated mathematical objects are generated, and the reader is asked whether a certain statement for the randomly generated objects is satisfied. Figure 2 shows an example in this regard.

(2.3) Analogue to section (1.1.1), but not static as in section (2.1), random graphics are generated, and the reader is asked whether a certain statement for the randomly generated objects represented in the graphics is satisfied. Figure 3 shows an example in this regard.

Note that the exercise shown in figure 3 is implemented by generating "invisible" pages in the document, one page for every position of the 4 needed arrows. Thus a 4-element set needs 16 invisible pages. The internal javascript engine of the pdf-document then generates a random choice of arrows which define a map $f\colon X \to X$, and makes corresponding pages "visible" to the reader when the problem is presented.

(2.4) Analogue to section (1.1.2), random objects are generated, and the reader is asked to enter the result of a computation involving the randomly generated objects (*e.g.*, the intersection of two randomly generated sets) into an answerfield. Figure 4 shows an example in this regard.

Observe that in this example, infinitely many correct ways to write the solution exist. Some of the javascript-code which is embedded in the pdf-document and checks the solution of the above problem (or, *e.g.*, the solution of a similar problem such as the computation of the power set of a small set) is based upon a grammar which specifies all correct solutions. The actual code which checks the solution is then generated via jison [10] (*i.e.*, "javascript-bison", compare [3]; compare also [13] in regard to the UNIX tools lex and yacc).

(2.5) Analogue to section (1.1.5), carefully crafted lines of a proof are presented in a randomly generated, permuted order, and the reader has to return the proper order of the lines of the proof. Figure 5 shows an example in this regard.

---

[7] All figures of this document are collected in an appendix at the end of the paper.

Obviously, it makes no sense to repeat this randomised exercise, once it has been successfully completed. However, if the document is studied again after a longer period of time, then a newly generated, random order of the lines in the proof *may* require the reader to read and rethink the mathematical formulas in the proof with greater care than in the situation of a once-fixed random order which may be easily memorised. Furthermore (compare section 3), observe that if such an interactive teaching document provides feedback to a central host which registers and administers the use of such documents as described in this work, *e.g.*, as part of a university course, then it is desirable that different students a presented with (most likely) different versions of the problem to suppress at least blind copying of solutions.

Note that the exercise is implemented by generating "invisible" pages in the document, one page for every position of every line of the proof. Thus a 20 line proof generates 400 invisible pages. The internal javascript engine of the pdf-document then generates a random permutation, and makes corresponding pages "visible" to the reader when the problem is presented.

The techniques used to implement problem types (2.1)-(2.5) have also be used to implement multiple-choice type exercises where, *e.g.*, the reader has to associate a randomly selected formula (from a finite family of formulas memorised within the pdf-document) with one of several displayed graphs in the document. This is a variation of problem type (2.2) where one of several buttons can be pressed to answer a question rather than only one out of two.

Altogether, the implemented interactive randomised exercises which predominately implement rather elementary mathematical problems provide nevertheless a vivid interactive learning environment, in particular, for the novice learner of mathematics.

This author has sought the feedback of some children, and also a few students of The University of Aizu in regard to the manuscript [21]. Children seem to like the elementary interactive portions of [21] and wish, *e.g.*, to compute set-intersections or set-unions repeatedly as if they were playing a game on the computer screen. Students of The University of Aizu expressed a positive attitude towards the overall concept of [21] as teaching material for fundamental content of their studies.

3   Extending to Learning Management Systems with Distributed Computing Load — An Outlook

We can (in future work) go beyond the scope of the documents and embedded computational abilities described in section 2 and take advantage of the ability of pdf-documents to communicate with dedicated machines on the internet. Thus, the results of evaluations which an interactive document produces can also be emailed or otherwise communicated back [8, p. 15] to an administrative server where a learning management system (*e.g.*, Moodle) is installed. Alternatively, all remaining, mostly administrative features of the UNEIM/MAT system as outlined in section 1 could, in principle, be reimplemented to run on a dedicated host for such a decentralised approach. What would be achieved this way (either recreating the UNEIM/MAT system, or connecting to another established learning management system) is a distributed computational load within the overall system[8], and a reduced need for communication via the internet. In addition, the user (*i.e.*, student) of such a system can work offline and still use interactive features of the overall system to a large degree.

Altogether, a decentralised, load-sharing approach to learning management systems as outlined above seems to be significantly advantageous, in particular, in regions where internet access is still relatively expensive (compare [19]).

Conclusion

In this work, we have explicitly shown how certain interactive features of learning management systems, *i.e.*, quizzes and certain types of exercises including evaluations and feedback, can be incorporated into LATEX-generated pdf-documents via embedded javascript-code. Such an approach to LATEX-generated interactive teaching documents is particularly advantageous, if the teaching of mathematics is considered. The embedded computational abilities of the enhanced pdf-documents which are discussed in this work allow within one session of use of the document for the repeated creation and presentation of exercises, if the latter are designed to contain components that are randomly generated or randomly selected from a limited family of predesigned choices stored within the document. This enables the user/student to use the embedded exercises in the described interactive teaching documents repeatedly for drill purposes.

Furthermore, the approach to the concept of a learning management system where interactive teaching documents are part of the system have been shown to balance the computational load between user side and server/administrative side while reducing the need for communication over the internet, and allowing the user independent offline use of larger portions of such a system.

Acknowledgements. The author would like to thank B. Katayama, the author of [11], for his programming work which provides a first version of functionality to the exercises in the interactive teaching document project "Set Theory" [21]. It is unfortunate that Mr. Katayama had to leave the project a significant amount of time before completion. The author would also like to thank P.-A. Fayolle for comments and discussions in regard to various aspects of this project.

References

1. PDF Reference, sixth edition. Adobe Corp. (2015),
   http://wwwimages.adobe.com/content/dam/Adobe/en/devnet/
   pdf/pdfs/pdf_reference_1-7.pdf

---

[8] Evaluations and most user-dedicated responses are generated on the user's side, while mostly administrative tasks are performed on the server/instructor/institution's side.

2. PDF Reference and Adobe Extensions to the PDF Specification. Adobe Corp (2015), `http://www.adobe.com/devnet/pdf/pdf_reference.html`
3. Bison. Free Software Foundation, `http://www.gnu.org/software/bison/`
4. Blackboard Website, `http://www.blackboard.com`
5. Blackboard: Tests, Surveys, and Pools, `https://help.blackboard.com/en-us-pro/Learn/9.1_SP_12_and_SP_13/Instructor/100_Tests_Surveys_Pools`
6. Question Types in Blackboard. iSolutions - the University IT Professional Service, University of Southampton, UK, `http://www.southampton.ac.uk/isolutions/computing/elearn/blackboard/questiontypes.html`
7. Date, C.J.: An Introduction to Database Systems. Addison Wesley (1995)
8. "Instructional Development and Authoring" Lab (IDEA), N. Dvoracek (Director/Interim CIO). Creating Files and Forms in Adobe Acrobat. University of Wisconsin-Oshkosh, USA, `http://idea.uwosh.edu/nick/Creatingfilesinacrobatv7.pdf`
9. The GNU General Public License (v3.0). Free Software Foundation, `http://www.gnu.org/copyleft/gpl.html`
10. Jison (javascript parser generator), `http://zaach.github.io/jison/docs/`
11. Katayama, B.: Interactive Pdf Documents for Computer-Based Education Using Javascript Inserted Into LaTeX. M.S. Thesis (supervised by L.M. Schmitt), The University of Aizu, Japan (2014)
12. Kawai, M.: A Development Environment for Randomized Quizzes within a Unix-based Learning Management System. B.S. Thesis (supervised by L.M. Schmitt), The University of Aizu, Japan (2008)
13. Kernighan, B.W., Pike, R.: The Unix Programming Environment. Prentice-Hall (1984)
14. Kuráňová, S.: Interactive PDF Documents in Math Education Focused on Tests for Differential Equations. Masaryk University, Czech Republic, `http://math.unipa.it/grim/21_project/Kuranova347-352.pdf`
15. LearnDash Co. 20 Most Popular Learning Management Systems, `http://www.learndash.com/20-most-popular-learning-management-systems-infographic/`
16. Moodle Website, `https://moodle.org`
17. Quizzes in Moodle. LSE — Learning Technology and Innovation, `http://lti.lse.ac.uk/moodle/quizzes.php`
18. Parker, T.: Form Submit / eMail Demystified, `https://acrobatusers.com/tutorials/form-submit-e-mail-demystified`
19. Smith, D.: Africa Calling: Mobile Phone Usage Sees Record Rise After Huge Investment. The Guardian, Manchester, UK (2009), `http://www.guardian.co.uk/technology/2009/oct/22/africa-mobile-phone-usage-rise`
20. Schmitt, L.M.: Email-Based Mathematics Teaching Tools. International Transactions on eLearning and Usability, The University of Aizu, Japan, Center for Language Research, 1–7 (2010), `http://clrweb.u-aizu.ac.jp/itelu/files/2011/05/2010-21.pdf`
21. Schmitt, L.M.: Set Theory — An Interactive Introduction for the Mathematical Layman. The University of Aizu, Japan, draft document, 1-68 (2015)
22. Schmitt, L.M., Christianson, K.T.: Pedagogical Aspects of a UNIX-based Network Management System for English Instruction. System 26, 567–589 (1998)
23. Story, D.P.: AcroTeX: Acrobat and TeX team up. TUGboat, 20, 196-201 (1999), `https://www.tug.org/TUGboat/tb20-3/tb64story.pdf`

24. Story, D.P.: Exerquiz & AcroTeX, Packages for Including Special Effects in Pdf Documents, using TeX and LaTeX. Department of Mathematics and Computer Science. University of Akron, http://www.math.uakron.edu/dpstory/webeq.html

25. Story, D.P.: Techniques of Introducing Document-level JavaScript into a Pdf File from LaTeX Source. TUGboat, 22, 161–167 (2001), http://www.tug.org/TUGboat/tb22-3/tb72story.pdf

26. Webb, J.H.M.: 121 Common Mistakes of Japanese Students of English (revised edn.). The Japan Times, Tokyo (1992)

27. Wikipedia, Blackboard Learning System, https://en.wikipedia.org/wiki/Blackboard_Learning_System

28. Wikipedia, List of Learning Management Systems, https://en.wikipedia.org/wiki/List_of_learning_management_systems

29. Wikipedia. Moodle, https://en.wikipedia.org/wiki/Moodle

30. Wolfram, S.: Mathematica: A System for Doing Mathematics by Computer (second edition). Addison-Wesley (1990)

Appendix: Collection of Figures

1.11 Exercise. (the empty set as an element in finite sets)

(a) Is $\emptyset \in \{0,1,2,3,4,5,6,7,8,9\}$?

| Yes | | No | Wrong.

Comment: The empty set $\emptyset$ cannot be found in the set of digits $\{0,1,2,3,4,5,6,7,8,9\}$.

(b) Is $\emptyset \in \{\emptyset,1,2,3,4,5,6,7,8,9\}$?

| Yes | | No |

(c) Is $\emptyset \in \{\{\,\},1,2,3,4,5,6,7,8,9\}$?

| Yes | | No |

Fig. 1. Excerpt from [21]. Exercise with yes/no type of questions. Answer is given by the reader via clicking the yes- or the no-button. After such a click, an evaluation and comment possibly including formulas appear in the document.

2.5 Exercise. (subset relation between finite sets) Let the finite sets S and X be given by the following equations:

- $S = \{2,9,1,4\}$

- $X = \{7,2,4,1,8,3,9,5,6\}$

- Is $S \subseteq X$?

| Yes | | No |

Fig. 2. Excerpt from [21]. Exercise with yes/no type of questions about randomly generated objects (here sets S and X). Answer is given by the reader via clicking the yes- or the no-button. After such a click, an evaluation appears in the document, and a new version of the problem is displayed.

15.3 Exercise. (bijective maps on a finite set)   Let $X = \{\bullet, \circ, \circ, \bullet\}$, *i.e.*, the four-element set of distinctly coloured bullets $\bullet$, $\circ$, $\circ$ and $\bullet$ depicted in two copies of identical boxes in figure 15.3.

In this exercise, a map $f: X \to X$ is given by (or represented by) the collection of black arrows from left to right in figure 15.3.

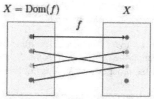

Figure 15.3: a map f on a 4-element set represented by 4 arrows from left to right

- Is the map f shown in figure 15.3 bijective?

  [Yes]   [No]

Correct. You may try this task again

Fig. 3. Excerpt from [21]. Exercise with yes/no type of questions about randomly generated graphics (here representing a map f on a set X). Answer is given by the reader via clicking the yes- or the no-button. After such a click, an evaluation appears in the document, and a new version of the problem is displayed.

4.3 Exercise. (intersection of finite sets)   Let X and Y be given by the following equations:

- $X = \{5, 3, 1, 9, 4, 6, 7\}$
- $Y = \{9, 3, 8, 6, 5, 7, 4\}$
- Determine $X \cap Y$

Enter your solution to the problem in the input-field below. Your input (*i.e.*, answer) should be similar to the string $\{\, 3, a, 1, 2, b, c \,\}$

$X \cap Y =$ [   ]

Fig. 4. Excerpt from [21]. Exercise to compute the intersection of two randomly generated sets X and Y. Answer is given by the reader via typing into an answer-field. After the answer has been provided, the internal javascript engine of the document evaluates the entry, and the result of this evaluation appears in the document. In addition, a new version of the problem is displayed

8.7 Proposition. (on the union of two disjoint finite sets) Let S and N be disjoint finite sets, and let X be given by $X = S \cup N$. Then $S = X \backslash N$ and $N = X \backslash S$.

PROOF (exercise): The lines of the proof of proposition 8.7 are presented in random order below. As an exercise, the reader is supposed to find the proper order of the lines of the proof.

(1) Finally, observe that the situation (or content) of proposition 8.7 is completely symmetric in the symbols S and N. To obtain $N = X \backslash S$, one can therefore simply exchange S and N in "$S = X \backslash N$" which is already proved.   ■

(2) Next, we show $X \backslash N \subseteq S$. If $x \in X \backslash N$, then again, according to definition 7.1, we have $x \in X$ and $x \notin N$. This implies $x \in S$, since $X = S \cup N$, and x must be somewhere. Hence, $X \backslash N \subseteq S$, and the proof of $S = X \backslash N$ is finished.

(3) To show $S \subseteq X \backslash N$, we start with an arbitrary element $x \in S \subseteq S \cup N = X$. Then, $x \in X$ and $x \notin N$, since the disjoint sets S and N have no common elements according to definition 6.1.

(4) By definition 7.1, $x \in X$ and $x \notin N$ together mean exactly that $x \in X \backslash N$. Hence, we have completed the proof of the statement $S \subseteq X \backslash N$.

(5) We begin the proof by showing $S = X \backslash N$, i.e., we prove $S \subseteq X \backslash N$, and then $X \backslash N \subseteq S$ according to definition 1.13 and proposition 2.6.

TASK: Rearrange lines 1-5 in proper order such that this yields a correct proof of proposition 8.7.

Enter your solution to the problem in the input-field below. Your input (i.e., answer) should be similar to the string 5, 4, 3, 2, 1

Fig. 5. Excerpt from [21]. Exercise with randomly generated order of the lines of a proof. Answer is given by the reader via typing the proper order of lines into an answer-field. After the answer has been provided, the first error (if existent) in the answer is marked, and the reader can rethink and update his solution.

Author Index